Chasing
Black
Holes

An Insider's View
of a Space Astronomy Mission

Hale Bradt

Massachusetts Institute of Technology
Cambridge, MA
bradt@mit.edu

V

VAN
DORN
BOOKS

Also by Hale Bradt

Cover (front and back): Artist's conception of an X-ray binary star system, one star of which can be either a neutron star or a black hole, which features an accretion disk of circulating X-ray emitting gases and also emerging jets of radiating particles. The inset is the author's MIT science team. See captions to Figures 21 and 22. [Credits: NASA/Dana Berry, Arlyn Hertz, Lisa Carta]

Opposite title page: Liftoff of XTE. See caption to Fig. 20.

In memory of

Bruno B. Rossi
(1905–1993)

Jan van Paradijs
(1946–1999)

Chasing Black Holes: An Insider's View of a Space Mission
by Hale Bradt © 2021

Based on a talk given at the Symposium,
"Exploring Physics with Neutron Stars,"
organized by Dimitrios Psaltis in honor
of the 65th birthday of Frederick K. Lamb,
Tucson, Arizona, 18 Nov. 2010

**VAN
DORN
BOOKS**

www.vandornbooks.com
info@vandornbooks.com

Editor: Frances B. King
Cover and Interior: Lisa Carta Design

Library of Congress Control Number: 2021902898

ISBN 978-0-9966939-5-0
SCIENCE / Physics / Astrophysics
SCIENCE / Space Science / Astronomy
BIOGRAPHY & AUTOBIOGRAPHY / Science & Technology

CONTENTS

ILLUSTRATIONS

Preface

I RETIRED FROM THE MIT FACULTY in 2001 and had to reduce my office footprint because I was to share the office with another faculty retiree. As I sorted through files and documents, many of which I threw out, I came across records relating to the orbiting X-ray astronomy observatory, the *Rossi X-ray Timing Explorer* (RXTE), a NASA space mission that I had worked on for many years. At that time, it had completed almost six years of its 16 years of productive observations in orbit. It occurred to me that there was a story therein, namely how one brings such a mission from concept to orbit, a task that took 21 years! I intended to donate the documents to the MIT Archives or Museum for the historical record. In the meantime, I left this pile of documents, about a foot high, on top of one of my file cabinets with the intention of creating a brief overview that would accompany it to the archives.

Nine years later (in 2010), that pile still lay on that file cabinet untouched when a colleague at the University of Arizona, Dimitrios Psaltis, called and asked me to give a talk on the selling of the RXTE mission. "Why me and why that?" I asked.

"Well," he said, "This symposium is to honor Fred Lamb on the occasion of his 65th birthday, and he was instrumental in helping bring about the RXTE mission." He went on to say, "I recall that, at a conference in Rome, over dinner or lunch, you told me all about the adventures of getting RXTE into orbit, and a lot of it was very funny. That's why I thought it would be appropriate for the banquet talk at Fred's birthday symposium." I was not sure it was all that funny, but I agreed, realizing this was the goad I needed to address that pile of untouched documents.

Procrastinating as usual, I finally got around to looking at them a few days before leaving for the meeting, which was held in Tucson, Arizona, at the Desert Museum. I had never kept a proper notebook of the events leading to the RXTE launch, but these documents brought back memories and pinned down some of the dates and events. Conversations with Fred Lamb and Rick Rothschild at the meeting brought more clarity to the story. At the banquet, I recounted the tale to the 40 or so participants. I went on for an unconscionable two hours(!), but nobody seemed to mind as we had nothing else to do until our bus to the hotel arrived. I finished up just as it arrived.

Dimitri urged me to write up the talk, which I did, beginning on the flight home, knowing that, if I didn't, I probably never would. I have revisited it several times since, as I learned more from my colleagues and my own investigations.

After more procrastination due in part to its candid description of interpersonal issues, I am now, finally, making it available here to my fellow high-energy astrophysicists and to interested lay persons. I have attempted to make it accessible to the latter, with only modest excursions into technical details.

Bringing about a mission such as RXTE is a complicated business and in many respects is a struggle. Bureaucratic, financial, and technical problems that at times seem overwhelming must be solved, and, in the process, the varied interests of individuals and organizations come to the fore. We at MIT were deeply involved as were the managers, engineers, and scientists at NASA's Goddard Space Flight Center, at the University of California at San Diego, and at NASA Headquarters.

When there were difficult issues where perspectives differed, we at MIT usually considered ourselves to be the "good guys" and the others, not "evil," but possibly a bit misguided in their understanding. The view from the other side could have been similar, or possibly more balanced than ours. At times the rhetoric and emotions would become rather warm, but we all wanted RXTE to fly and were working toward that goal. Many of us were friends before the RXTE project and remained so during and after it.

In recounting these events, I do try to present others' views a little more clearly than I did at the time. Nevertheless, my MIT bias is sure to show. Memories can be faulty so aspects of this may be forgotten or misremembered. I have tried to verify events by consulting participants and by referencing existing documents. However, at this final writing, during the Covid-19 pandemic, my collection of RXTE records is at the MIT Museum and not now available to me, as it was for my earliest drafts.

This is very much my personal story; it is not an objective history. Others surely would tell it differently. Any errors in this document belong on my doorstep.

For those with an engineering viewpoint, the ultimate goal was to put an operating spacecraft and science instruments safely into orbit. For scientists like myself, the goal was to get data from black holes and neutron stars with a new powerful instrument. But along the way, there was substantial satisfaction in each small step that brought us closer to the final goal. This is largely the story of these steps from my perspective. The science results presented also reflect that perspective. This is a behind-the-scenes story of a space mission that is rarely told.

And, yes, four years after the Lamb Symposium, I finally donated those RXTE documents in my office to the MIT Museum. They were accompanied by prototypes of the instruments we created for RXTE, the All-Sky Monitor and the Experiment Data System. Both are on loan to the MIT Kavli Institute, where they will soon be on display.

Hale Bradt
Salem, MA, January 2021

1
Introduction

THE *ROSSI X-RAY TIMING EXPLORER* (RXTE) was known as the X-ray Timing Explorer (XTE) during its development; it was christened with the Rossi name when it was launched into space on December 30, 1995. Bruno Rossi (1905–93) was a renowned scientist, one of the founders of X-ray astronomy, and a long-time professor of physics at MIT.

The RXTE was one of NASA's small "low-cost" Explorer series of satellites. But it was not small in size (6 ft × 6 ft × 18 ft), weight (7055 lbs), or cost (close to $200 million in 1989 dollars). This was a lot of money, but not as much as the most ambitious NASA space missions; the *Hubble* and *Chandra* observatories each cost about ten times that. The overall mission as it was finally configured is described by me and the other principal investigators, Jean Swank and Richard Rothschild, in a 1993 paper.[1]

The RXTE (Figs. 1 and 2) was designed to study, for the most part, compact objects (neutron stars and black holes), which are often strong emitters of X-rays. Such objects have extremely

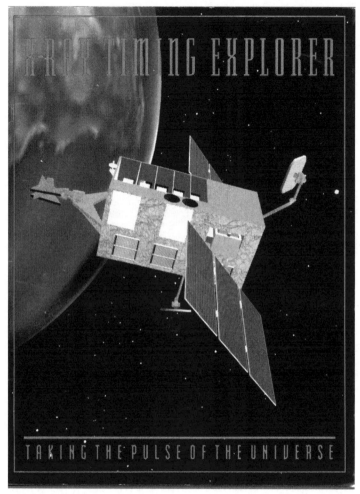

Figure 1: Rendering of XTE in orbit

Artistic rendering of the X-ray Timing Explorer in orbit on the cover of the brochure we distributed at the January 1992 meeting of the American Astronomical Society. The lower label, "Taking the Pulse of the Universe," was Rick Rothschild's creation. The MIT instrument, the All-Sky Monitor, is on the pedestal at the far left.

strong gravitational fields, which are often associated with very hot gases, gases so hot that they emit X-rays copiously. The gravity is so strong near black holes and neutron stars that the X-ray emitting gases would move distances comparable to the object's "size" in only a few milliseconds. Thus, one might expect variations in the X-ray intensity on such time scales. Those variations would expose the physics taking place in the immediate vicinity of the black hole or neutron star, for example the time it takes matter to orbit the compact object. RXTE carried three science instruments to address such phenomena.

The first instrument, the Proportional Counter Array (PCA), was the heart of RXTE. It was designed to study such variability. It featured a large detection area so it could collect as many X-rays as possible in a short time. If one wants to study variability on the time scale of milliseconds, one must be able to collect at least a few X-rays in a millisecond for the brightest sources. The instrument consisted of five large proportional counters with a total effective area of 6250 cm² (2/3 of a square meter or 3/4 of square yard). A proportional counter is a sealed box of gas wherein an X-ray is converted to an electric pulse of magnitude roughly proportional to the X-ray energy (color).

Stephen Holt of Goddard Space Flight Center was the initial principal investigator for the PCA. Jean Swank replaced him in that position in 1990 when Steve became chief of the space sciences at Goddard. Peter Serlemitsos of their science group had developed this type of counter for earlier space experiments. His colleagues Keith Jahoda, Will Zhang, and Chuck Glasser led the development of the larger detectors needed for RXTE.

The PCA did not take pretty photographs of objects in the X-ray sky such as those produced by the *Einstein* and *Chandra* missions. They featured reflecting X-ray telescopes

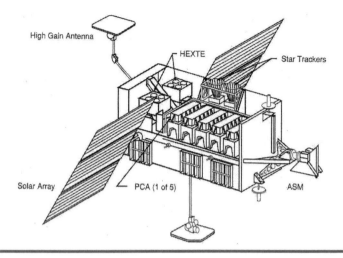

High Gain Antenna

HEXTE

Star Trackers

Solar Array

PCA (1 of 5)

ASM

Figure 2: Arrangement of instruments in spacecraft

Sketch showing the RXTE final configuration, but without the thermal-control coverings, showing the three science instruments: the Proportional Counter Array (Goddard Space Flight Center), the All-Sky Monitor (MIT), and the High-Energy X-ray Timing Experiment (U. Calif. San Diego). The dimensions of the actual spacecraft were 6 ft × 6 ft × 18 ft, excluding the extended antennas and solar arrays. It weighed 7055 lbs (3.5 tons). The MIT Experiment Data System is not visible.

that focus the X-rays to an actual image. In contrast, the XTE PCA could be called a large "light bucket" that could gather in large numbers of X-rays every second, in part because it was not handicapped by the lower throughput of focusing X-ray optics. To achieve sensitivity to millisecond variability, this light bucket also needed fast detectors, precision clocking, and sophisticated event processing to transmit the information to the ground. The PCA also provided rough spectral information, namely the *colors* of the detected X-rays.

The PCA carried metallic honeycomb-like collimators that limited its view of the sky to the one-degree patch of sky toward which the instrument would be pointed. This made possible the study of just one X-ray source at a time; the brighter X-ray sources are mostly spaced much farther apart. The one-degree collimation meant the *angular resolution* of the instrument was one degree. This is quite poor compared to that of the human eye (about an arc minute) or to ground-based optical telescopes (about one arc second). In short, meaningful images were unattainable by RXTE.

The focusing X-ray missions *Einstein* and *Chandra* were much more sensitive. They had greater angular resolution (down to about one arcsecond) and detectors with low backgrounds. Hence, they could detect much fainter sources. However, they could not accumulate large numbers of X-rays in short time periods because of their limited effective areas, nor could they reach to the higher-energy X-rays that are important for modeling the emission processes. RXTE was a *timing* mission; it complemented the *imaging* missions.

RXTE was able to follow up on the previous three decades of temporal studies of X-ray binary star systems with their rich yield of results on pulsars, transients, and nuclear outbursts. It did so by exploring aspects with greater observing power than did its predecessors, due to its large collecting area and ability to reach millisecond time scales. But regrettably, it did not provide colorful pictures of the sky like you see in *Sky and Telescope* and occasionally in the *New York Times*.

The second instrument on RXTE was the High Energy X-ray Timing Experiment (HEXTE). It extended the energy range of detected X-rays from the upper limit of the PCA (60 keV) to 200 keV and was designed and built by the University of California at San Diego. Richard (Rick) Rothschild was the principal investigator. It, like the PCA, had a small field of view which was coaligned with the PCA view direction. It enhanced the power of the PCA by observing high-energy X-rays, the X-ray *colors* the PCA could not reach. Rick was ably assisted by scientists Biff Heindl and Paul Hink, project manager Ed Stephan, and techs Fred Duttweiler, Phillip LeBlanc, Ed Stephan, and Pete James.

The third scientific instrument carried on RXTE was the All-Sky Monitor (ASM), designed and built at MIT. X-ray stars in the sky are highly variable in intensity, unlike the visible stars one sees at night. Some actually disappear altogether and then, in some cases, reappear weeks or years later. The PCA and HEXTE instruments "see" only a tiny piece of the sky (one-square-degree) at any one time and would be blind to interesting or unusual behavior elsewhere on the sky; there are about 40,000 square degrees over the entire sky.

The RXTE thus needed "eyes" to watch the entire sky and alert it to interesting activity for the PCA and HEXTE to study. If such an object were found by the ASM, the entire satellite could then be reoriented to focus the PCA on it. The ASM kept a watch on the sky by scanning over most of it once every 90 minutes. It was of course less sensitive to fainter sources, because of its lesser collecting area and because it spent less time observing each part of the sky. It was the perfect complement to the powerful PCA/HEXTE, the Cyclops of RXTE.

The importance of an All-Sky Monitor had become evident with the great success of the ASM on the British *Ariel-V* satellite (launched 1974). That experiment consisted of a pair of pinhole X-ray cameras built at Goddard. Steve Holt was the principal investigator, and his student Lou Kaluzienski produced paper after paper of interesting results. (Before I met Lou, I thought maybe that Steve had constructed a paper-writing machine he dubbed "Kaluzienski.") MIT's SAS-3 satellite experiment (launched in

1975), with long narrow fields of view, also detected changes in source brightness and discovered previously unknown sources. It too illustrated the value of a sky monitor.

I was the initial principal investigator of the ASM on RXTE. After my retirement in 2001, my MIT colleague Alan Levine took on that role. The ASM is described in some detail in a paper[2] that Al published with others of our group. The MIT scientists involved throughout the entire RXTE program were Levine, Edward Morgan, and Ronald Remillard. The latter two were former PhD students of mine, as was Bill Mayer, our project manager. Bill managed the entire MIT portion of the RXTE project and was ably assisted by project engineer Bob Goeke and quality assurance manager Brian Klatt.

Our group was also responsible for providing the on-board computer that processed the data from the PCA and the ASM, namely the Experiment Data System (EDS). It met the challenges of handling with flexibility the very high data rates required for the study of matter near black holes and neutron stars. Despite a superb telemetry link to the ground, the very high data rates of the brightest sources could overwhelm the link; it was thus necessary to compress the raw data. We strove to provide observers the greatest possible flexibility within these constraints. The design required the creativity and innovation typical of a scientific instrument, but the EDS did its work behind the scenes.

RXTE was launched on December 30, 1995 from Kennedy Space Center at Cape Canaveral, Florida. It produced data for 16 full years, into the first days of 2012. On January 5, NASA's funding of its operations came to an end, and it was decommissioned. It then silently orbited the Earth until its reentry into the atmosphere on April 30, 2018, after spending exactly 22 and

TABLE 1: ACRONYMS AND PRINCIPAL INVESTIGATORS	
Mission before launch (Dec. 30, 1995)	**Mission after launch**
XTE *X-ray Timing Explorer*	**RXTE** *Rossi X-ray Timing Explorer*
Institution	**Instrument and Principal Investigator(s)**
GSFC aka Goddard Goddard Space Flight Center, NASA	**PCA** Proportional Counter Array Stephen Holt (1982), Jean Swank (1990),
MIT Massachusetts Institute of Technology	**ASM** and **EDS** All-Sky Monitor; Experiment Data System Hale Bradt (1982), Alan Levine (2001)
UCSD University of California at San Diego	**HEXTE** High Energy X-ray Timing Experiment Richard Rothschild (1982)

1/3 years in orbit. It produced a wealth of results reported in over 2000 scientific papers in refereed journals, some of which are described in Chapter VII.

Acronyms are the province of nerds and should be minimized in popular works. Here, I beg the reader's forbearance in learning an essential few, namely those for the mission, those for the three institutions providing instruments, and those for the instruments provided by those institutions, as shown in Table 1. A number of other acronyms are used for efficiency, but only shortly after their definition. In other instances, the acronym can be accepted simply as a name, e.g., NASA, MIT. Appendix I carries a list of the acronyms and abbreviations found herein.

For this story, it is important to understand the different NASA entities with which we had to deal. The project office at NASA's Goddard Space Flight Center managed the entire RXTE program for NASA; it monitored the progress of all three instrument teams and distributed money to them as needed.

The scientific group at Goddard that provided the PCA was a completely separate entity. Another group at Goddard built and integrated the spacecraft.

The three groups at Goddard were all located on its campus in Greenbelt, Maryland, but in different buildings. Greenbelt is a suburb of Washington, DC. NASA Headquarters is 16 miles away in downtown Washington. It was the ultimate source that allocated money to the project office and determined whether or not there would even be an RXTE mission. The individuals in these several offices were some of the many players involved in the scientific, political, bureaucratic, and technical dance we are about to describe.

This account is primarily chronological and extends over the 21 pre-launch years. Adding the 16-year duration of post-launch science operations makes that 37 years!

II
Early Concepts
(1974–1979)

The scene in Cambridge, Massachusetts

THIS STORY BEGINS in the early 1970s. I was in my early forties, full of energy, and a faculty member at MIT in Bruno Rossi's cosmic-ray group, where I was eagerly pursuing the nascent field of celestial X-ray astronomy. The field had begun in 1949 with the discovery of X-rays from the sun by means of sounding rocket flights carried out by Herbert Friedman's group at the Naval Research Laboratory (NRL). The rockets—similar to the German V-2 rockets that bombarded London in World War II—carried their detectors above most of the atmosphere, a necessity because the atmosphere is opaque to X-rays coming from distant celestial sources. Each flight provided only a few minutes of observation time before the rocket fell back to Earth. The landings were cushioned by a parachute, which preserved the instruments and payload structure for use another day.

Although the sun was quite bright in X-rays and hence easily detected, the energy flux in X-rays was minuscule compared to that of visible light. It was generally considered hopeless to look for sources from elsewhere in the sky because the closest stars— being distant suns—would be undetectably faint in X-rays at their great distances.

That realization did not deter a group at AS&E, a small firm near MIT in Cambridge, from taking a look. AS&E had been founded by Martin Annis and several MIT faculty, including the renowned Bruno Rossi, the leader of MIT's cosmic-ray group. Bruno and AS&E scientist Riccardo Giacconi were cognizant of energetic events in the cosmos that were not well understood, such as cosmic rays, supernovae, and quasars. They also knew that celestial X-rays could propagate freely through the gases and grains of interstellar space. Furthermore, they knew the discovery potential of investigations in previously unexplored areas. "If no has ever looked there, discoveries are likely." Think of Galileo and the telescope and Columbus with his ships. Since no one had searched the sky more broadly for X-rays, they proposed funding for a rocket experiment that could take such a look.[3]

With a 1962 rocket flight, Giacconi, Rossi, and their collaborators discovered an intense unexplained source of X-rays from the constellation Scorpius.[4] Its nature was a complete mystery. This inspired NRL and MIT as well as a few other groups in the US and abroad to prepare and carry out their own experiments to probe this new phenomenon. Thus was opened a whole new field of *celestial* (as distinct from solar) X-ray astronomy.

At MIT, in the following years, George Clark, Gordon Garmire, and Minoru Oda collaborated with AS&E scientists in their ongoing sounding rocket program, while George initiated an MIT program in X-ray astronomy from high-altitude balloons, a program later taken over by Walter Lewin. In 1966, I joined in with Minoru on an AS&E flight that refined the celestial location of the bright source in Scorpius, Sco X-1, with sufficient precision (arc minutes) to lead to its identification with a bluish optical star, which was later found to be a two-star system containing a neutron-star.[5] A rocket program at MIT was begun by our group in 1966, and I was the principal investigator of that program. Saul Rappaport became its leader a few years later.

In the late 1960s, X-ray astronomers were just beginning to place instruments on orbiting satellites, which provided much longer observing times (one or more years) than balloons (hours) or sounding rockets (minutes). The first were "buses" that carried multiple experiments of different disciplines. The X-ray experiments had rather small apertures and could not be pointed at a single source; they typically scanned over large patches of the sky. Notable were *Orbiting Solar Observatories* #3 and #5 (OSO-3 and OSO-5). In 1971, an X-ray experiment provided by our group, led by George Clark for the NASA bus OSO-7, began collecting data.

The first NASA satellite dedicated solely to X-ray astronomy was proposed and carried out by the AS&E group under Riccardo Giacconi. It was *Small Astronomy Satellite* #1 (SAS-1), which was christened *Uhuru* upon its 1970 launch. It was operational until 1973 and was hugely successful. It demonstrated that the field of X-ray astronomy was as rich in phenomena as radio and optical astronomy.

As our story begins, George's experiment on OSO-7 was just, in 1974, completing its observations. At the same time, George Clark, Walter Lewin, and I were involved in the forthcoming large HEAO (*High Energy Astronomy Observatory*) satellites, the first (HEAO-1) to be launched in 1977 and the second, later named *Einstein,* in 1978. In addition, we were all, together with Herb Schnopper, immersed in the final preparations for the launch of the *Small Astronomy Satellite #3* (SAS-3), which would carry MIT experiments exclusively and be launched in 1975. It would be the first X-ray satellite that could point continuously, up to an hour, at a given X-ray star. It too would be a quite successful, as was its competitor for scientific glory, the British satellite *Ariel-V,* which flew at the same time with comparable capabilities.

In 1974, NASA was already planning for astronomy missions beyond these missions. It was in the midst of planning for the future *Hubble* mission for optical astronomy and had not overlooked the need for progress in other branches of astronomy. The release by NASA in 1974 of an Announcement of Opportunity opened a door for us; it was a request for proposals for a scientific mission in space astronomy. Even though at MIT we had not even launched SAS-3 nor seen our experiments on the HEAOs fly, we knew we could not let this opportunity for our more distant future, the 1980s, pass us by. The 1970s was a golden age of X-ray astronomy, and we wished to be involved in its continuation.

At that time, the feeling in the Rossi group was very much that we were a team and would share the labor and the benefits of our research efforts. The feeling was, "OK, we had better respond to this Announcement of Opportunity, so who has the time and energy to take it on?" That turned out to be me.

I wasn't alone in the effort to produce a proposal; other members of the group could and did contribute ideas as to the most productive approach. We had to extrapolate beyond the missions that would fly first when we didn't even know their findings! The Center for Space Research, the home of the Rossi group at MIT, had the financial and engineering infrastructure to support such proposals and the large NASA contracts that hopefully would follow. Our job as scientists was to come up with the scientific objectives and the instrumental concepts. As physicists with rocket and balloon and some satellite experience, our conceptions were practical and could actually be implemented. Our engineers would ensure this was so and would contribute their expertise to the proposal content. Our managers would contribute financial and schedule estimates.

LAXTE on a Scout, November 1974

The aforementioned NASA Announcement of Opportunity, was, for reasons I do not remember, issued in two parts, numbered 6 and 7. To X-ray astronomers with long memories they were known as AO 6/7. Jointly they requested proposals for *small* "Explorer" missions that would follow the series of three large HEAO satellites.

In response, we at MIT with the X-ray astronomy group at the University of Leicester in England submitted a proposal entitled *Large Area X-ray Timing Experiment (LAXTE) on a Scout Vehicle* (Fig. 3). I was the principal investigator (PI) and Ken Pounds of Leicester was the co-PI. It featured only one instrument, a large array of proportional counters—a precursor to RXTE's PCA. Unfortunately, the proposed payload lacked a sky monitor; the importance of studying unexpected activity in X-ray sources was not yet sufficiently appreciated. Nevertheless, with only one powerful instrument, it was beautiful in its simplicity.

The argument for a large detection area and pointed observations of X-ray sources was pretty obvious from the results of earlier observations with balloons, rockets, and the *Uhuru*

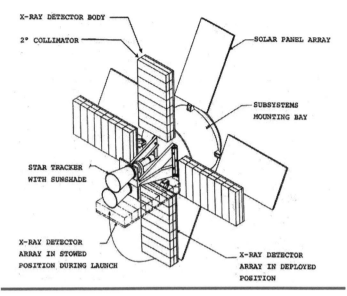

Figure 3: Large Area Timing Explorer (LAXTE)
Large Area X-ray Timing Explorer (LAXTE) from the 1974 MIT–University of Leicester proposal in response to the NASA 1974 Announcement of Opportunity, AO 6/7.

satellite. Larger collectors and more observing time with continuous pointing meant that an observer could seek out and study fainter objects and could follow them with time. The *Uhuru* satellite (launched in 1970) was showing the X-ray sky to be rich with objects emitting X-rays with highly variable intensities.

We were not the only proposers on AO 6/7 for X-ray instruments. Similar pointed X-ray experiments were proposed by groups at Goddard and the Harvard-Smithsonian Center for Astrophysics in Cambridge, MA. (The Smithsonian group, led by Riccardo Giacconi, had previously been at AS&E.) Both groups included friends from previous collaborations. Now they were our competitors!

This AO had a significant downside for X-ray astronomers.

They had long been the prima donnas of space astronomy because X-rays from space can be detected *only* from above the Earth's atmosphere. NASA Announcements in X-ray astronomy typically generated only a few proposals from the very few US institutions with programs in the field, and not all would respond to a given opportunity. Thus, X-ray astronomers could propose for a mission with a fair chance of it being accepted.

However, by the mid-1970s, scientists working in other branches of astronomy (microwave, infrared, and optical) began to wake up to the opportunities of space observations, even though some of their observations could be done fruitfully from the ground. For example, infrared astronomers were restricted to narrow frequency bands when doing astronomy through the atmosphere, and optical astronomers had to deal with the blurring of images by atmospheric turbulence. Space offered real advantages to these fields.

The AO 6/7 from NASA allowed for experiments in all wavebands. As a result, there was a plethora of proposals from all corners of the electromagnetic spectrum except the optical, which was well covered by the incipient *Hubble* program. The X-ray astronomers had real competition for a change. It was distressing, though, because the judging of the science merits from one waveband to another could be rather subjective. One might as well ask whether vegetables or fruits are more important. With no clear answers, the selections became more political, and here the X-ray astronomers were at a real disadvantage. "Hey, look at all the missions those X-ray guys have had!" (There were few women in the field in those days.)

It took some time for the AO 6/7 proposal selection process to play out. The preliminary result was neither a win nor a clear loss for us. NASA requested six different proposers of X and gamma-ray space missions to get together for a study of whether a viable single mission might be created with elements from their several proposals. The six proposers were thus sent off to carry out a joint study, each hoping to emerge alive.

Alive? Was it that serious? Yes, it was. Not only was one's own scientific future at stake but also the vitality of one's (probably small) research group; the funding from a successful outcome would cover grad students and technicians. Of course, if losing, you could do other things, but they might not be at the forefront of your field. The fork in one's career could well depend on the outcome of a committee's deliberations.

The chosen participants were Doyle Evans (Los Alamos Scientific Lab, LASL), Paul Gorenstein and Harvey Tananbaum (Smithsonian Astrophysical Observatory, SAO), Stephen Holt (Goddard), Seth Shulman (Naval Research Lab, NRL), and me (MIT). We had not yet lost the game; we were in extra innings!

ATREX, December 1976

In December 1976, our ad-hoc group submitted to NASA the result of its study. We proposed a viable mission (Astrophysics Transient Explorer; ATREX) that encompassed both X-ray and gamma-ray astronomy (Fig. 4). The two X-ray instruments were a narrow field of view, large-area array of detectors (a precursor to the XTE PCA) and a sky monitor patterned after the one launched on *Ariel-V* in 1974. The gamma instruments were a wide-field sensitive burst detector (essentially the BATSE experiment later flown on *Compton*) and a slit system for determining gamma-burst positions.

This collection of four instruments could be viewed as a bit of a "Christmas Tree," a derogatory term for missions that try to do too many things, rather than doing one or two of them well.

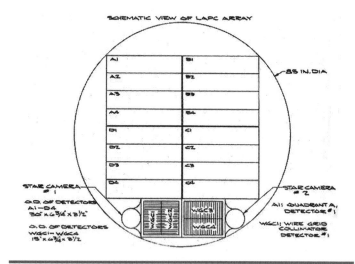

Figure 4: Astrophysics Transient Explorer (ATREX)
Astrophysics Transient Explorer, ATREX, the X-ray/gamma-ray mission proposed by the high-energy study group selected from the AO 6/7 proposals. The reader need not understand the details in the figure. This and Fig. 5 are shown solely to illustrate the reality of the failed proposals and the work—scientific and engineering—that went into preparing them.

However, that was not a fair description. ATREX addressed two major problems in astrophysics, the nature of neutron-star and black-hole systems and gamma-ray bursts in fundamentally new and powerful ways. In retrospect, it was a mission that could have jump-started by many years progress in these fields.

Unfortunately, the selection committees and NASA did not share our view of the importance of this mission, and it was not selected for flight. Instead, missions in other wavebands that later became quite well known and productive were selected: COBE (microwave), EUVE (extreme ultraviolet), and IRAS (infrared). These were all relatively small and inexpensive missions called Explorers. At the same time, more or less, a Great Observatories program was getting underway. It *eventually* led to the launch of the *Hubble* optical observatory in 1990, the *Compton* gamma-ray observatory in 1991, the *Chandra* X-ray observatory in 1999, and the *Spitzer Space Telescope* (infrared) in 2003. The Explorers were small potatoes compared to these.

The future beyond the HEAOs, in the 1980s, seemed barren indeed to us X-ray astronomers. Perhaps one more gambit was worth a try.

LAXTE on the Multi-mission Modular Spacecraft, November 1979

On November 17, 1977, we at MIT, in the wake of the AO 6/7 disaster (for X-ray astronomy), decided to take the bull by the horns and submit an *unsolicited* proposal to NASA for an X-ray timing mission. Since the previous multi-institutional proposals (LAXTE/Scout and ATREX) had failed, we chose a single-institution approach. A single involved institution would simplify management and reduce costs. We proposed (Fig. 5) a *Large Area X-ray Timing Experiment on the Multi-mission Modular Spacecraft*.

The Multi-mission Modular Spacecraft (MMS) had been developed by Goddard for the *Solar Maximum Mission* (SMM). It was designed to be recovered and serviced in the bay of the Space Shuttle while in orbit. The spacecraft would be put together with standard modules, which could be used for different missions. It would provide solar power, telemetry, and other services to payloads that could be mounted on it. Like a car assembly line, it could greatly reduce the per-unit costs, compared to creating customized spacecraft. (The first flight of the Shuttle occurred in 1981. In 1984, the SMM/MMS was actually retrieved and repaired in orbit with the Shuttle.)

We also proposed that the launch of LAXTE into orbit be accomplished with the Space Shuttle, because at that time—long before its first flight—it was being touted as a low-cost truck ride into space and a potential boon to scientific research. Once in orbit, the Shuttle would release the LAXTE/MMS into its own orbit. We further proposed that MIT do spacecraft integration and that one third of the observing time be allocated to guest (i.e., not MIT) observers. Guest programs were just coming into vogue; the data from early space missions had typically been analyzed only by the teams who built the observing instruments.

The main detector for LAXTE was again a large area (1 square meter) of proportional counters with a narrow field of view. It also included special detectors with slat collimators for periodic scanning the sky by means of spacecraft maneuvers, in other words, a sky monitor.

This was a complete proposal that represented a lot of scientific and engineering effort—spacecraft integration would be new to our laboratory. It may have seemed futile to make such an effort in the face of no announced opportunity by NASA. Needless to say, our colleagues at other institutions took a rather jaundiced view of this: "What a waste of effort!" But we knew the concept was a deserving one, and we were in an aggressive, fighting mood after the AO 6/7 debacle.

NASA simply ignored our proposal. It is possible that we got a nice refusal letter after some time, but I am not sure of that. Perhaps it helped the case for X-ray timing at NASA Headquarters in the long term, but maybe not.

I am reminded that MIT's Sam Ting (Nobel laureate) convinced the NASA Administrator (Daniel Goldin) in a direct personal presentation to fly his cosmic ray experiment,

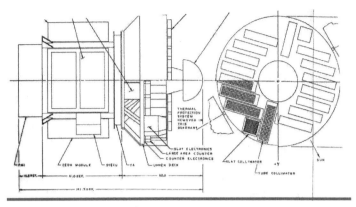

Figure 5: LAXTE on the MMS

LAXTE on the Multi-mission Modular Spacecraft (1977). This is from an unsolicited proposal from MIT to NASA that was simply ignored. Again, details are unimportant.

AMS, on the *International Space Station* in the absence of any announced opportunity. (At least, that is what I recall hearing. There is undoubtedly more to the story.) Could I have made such a pitch that carried any weight without the Nobel prize in my pocket? I doubt it, but perhaps I should have tried.

Thus, in 1978, the XTE concept was dead in the water. However, we X-ray astronomers at MIT were hardly bored. We were immersed in data pouring in from our experiments on SAS-3, HEAO-1, and the newly launched *Einstein*. But the view ahead was bleak. All of these missions were short lived, a few years at most.

The net result of all this effort was that it heralded more than a decade (1981–1995) of absolutely *no* sustained US X-ray astronomy observations aboard orbiting satellites. To pull off that complete shutdown of a productive field was no small feat on NASA's part. It was brought about principally by the infamous AO 6/7, severe overruns in the *Hubble* telescope program, and the Space Shuttle *Challenger* disaster. AO 6/7 was productive in that it led to spectacular results from COBE and IRAS, but better planning might have avoided completely shutting down US X-ray astronomy. Some rocket and balloon flights continued, but their contributions were specialized and limited.

In the 1970s, US X-ray astronomers had the use of the missions *Uhuru*, SAS-3, OSO-7, OSO-8, HEAO-1 and *Einstein*. X-ray astronomy had demonstrated that it had the full richness and potential of optical and radio astronomy. It could investigate with high statistics the strongest gravitational fields, the highest temperatures, the greatest densities, the highest particle energies, and the highest magnetic fields in objects as diverse as normal stars, neutron stars, stellar black holes, supernova remnants, and active galactic nuclei. It deserved support and needed space platforms to make further progress. Fortunately, as we shall see, the field did continue, but at a lower level, thanks to developing X-ray space programs in Europe and Japan.

Gorenstein points the way

The game then changed to one of US science politics, namely developing broad support for the mission concept in the scientific community with a high enough priority so it could get funded. NASA of course responds to the pressures it feels in the budgetary, political, and scientific spheres. All of that, of course, ultimately comes from us, the taxpayers and scientists. As Pogo said, "We have met the enemy and he is us."

It wasn't just a matter of selling an MIT interest. All X-ray astronomers had to support the project concept as good for the field as a whole. And then they would have to convince the rest of the astrophysics community. No one had to concede it to be more important than their own interests, just that it was worthy of inclusion of the mix of astronomy instruments in space. Success required an all-hands effort. If the concept of a timing mission was successful, no one would have a lock on building the scientific instruments; it would be an open competition.

The first steps were, of course, the multiple AO-6/7 proposals by us and others that gave the concept visibility in the community. The next significant step of which I am aware occurred shortly after the non-selection of ATREX. I was on a NASA advisory committee, the High Energy Astronomy Management Operations Working Group (HEAMOWG—yes, we pronounced the acronym as a multi-syllable word). It was directed by Albert Opp of NASA Headquarters who oversaw NASA's high-energy astronomy programs; he was the fountainhead, but not the source, of money for space research programs in X-ray astronomy. This committee had about a dozen scientist members who were active in X-ray, gamma-ray, and cosmic-ray research. It was intended to advise Dr. Opp on issues he faced. HEAMOWG was the lowest rung of NASA's scientific advisory system.

The HEAMOWG met shortly after the dismal AO 6/7 results were announced, probably in 1978. Dr. Opp needed our recommendations for space programs in high energy astronomy that NASA should address in the next round of proposal solicitations. On committees such as this, the members are supposed to represent their entire field fairly and not to push their own favorite projects. My baby, of course, was the pointed large-area X-ray observatory concept, but I dared not speak up lest it seem too self-serving. There was silence around the table, because everyone was probably dealing with the same conflicted feelings. It was awkward in the extreme; at least it seemed so to me.

Then, up spoke Paul Gorenstein of the Smithsonian group in his typically agreeable and thoughtful manner, "Well, many of the objectives we are interested in are to be addressed by accepted missions—gamma rays with *Compton* and focused X-rays with AXAF (later named *Chandra*)—but a very important one that is not being addressed is the study of X-ray sources with high time resolution, which requires large effective detecting area. I think we should propose that as our highest priority." Paul was an X-ray astronomer with broad scientific interests, so his words carried great weight in the committee, and, as I recall, the committee endorsed his view. I loved Paul for that. I had known him from his graduate-student days at MIT and later at AS&E in the early days of X-ray astronomy and had always liked him, as did most people. He was (and still is) so thoughtful and reasonable.

Our job was now reduced to convincing the community that such a mission was an important goal in X-ray astronomy, a goal that did not become easier when the *Einstein* mission (launched in 1978) began producing spectacular X-ray images of celestial objects. To some, that consigned non-focusing experiments, such as the large-area concept, to "old technology."

Such a view missed the point that the dominant aspect of X-ray astronomy had long been studies of temporal phenomena with broad-band spectra: pulsars, orbiting binaries, X-ray bursts, and transients (X-ray novae). It made no sense to stop such productive research simply because new techniques opened up different domains of study (high-resolution imaging and spectroscopy). One learns entirely different things from each.

Theorists to the rescue; Space Studies Board endorsement, April and August 1979

In 1979, Fred Lamb and David Pines, highly respected theorists at the University of Illinois, knowing the value of a timing mission to the physics of neutron stars, decided to do something about it. They organized a workshop in Washington, DC, on April 20–21, and edited the proceedings—papers written by the participants—entitled *Compact Galactic X-ray Sources: Current Status and Future Prospects*. This was the famous Orange Book that graced our bookshelves as a sort of bible for years (Fig. 6).

Fred reminded me that the attendees were required to construct a plan for a viable mission for inclusion in the proceedings and that the doors were locked (figuratively) until the mission concept was agreed upon and the writing justifying it was completed. (I was there but had forgotten those details.) The core mission was defined as a large-area pointed experiment (1–30 keV) and a sky monitor to track source behavior. The value of extended energy ranges, higher and lower, was stressed, but they were not part of the recommended core program.

This and other efforts led a few months later to endorsement by the highly visible committees of the National Academy of Sciences, specifically the Committee on Space Astronomy and Astrophysics (CSAA) and the Space Studies Board (SSB). Their report of August 1979, *A Strategy for Space Astronomy and Astrophysics for the 1980s*, devoted an entire two-sentence paragraph—on page 85 of a 113-page document—to the merits of an X-ray timing mission:

> Many objectives of high importance, however, cannot be met with [X-ray] telescopes. For example, the measurement of intensity variations over a wide range of time scales, which offers the potential of establishing

COMPACT GALACTIC X-RAY SOURCES

Edited by
Frederick Lamb
David Pines

Physics Department
University of Illinois
at Urbana-Champaign

Figure 6:

The Lamb/Pines Orange Book

Orange covered proceedings of the workshop organized by Fred K. Lamb (with tie) and David Pines, both of the University of Illinois Urbana-Champaign. This was the "Orange Book," which made the case for an X-ray Timing Explorer (XTE) mission to study compact galactic X-ray sources. Fred remained an enthusiastic supporter of the XTE mission throughout its entire life, including serving as chairman of its RXTE Users' Group, 1997–2008. [Photos: U. Illinois Urbana-Champaign] characteristics of neutron stars and black holes, needs to be incorporated into an Explorer mission of limited scope dedicated to timing measurements.

It wasn't much, but its importance was huge, because NASA pays close attention to the SSB. Since David Pines was then a member of the SSB, there is little doubt he had a hand in this.

Endorsement by HEAMOWG, sailing, and a lymphoma

A few months later, the HEAMOWG was convened by NASA's Al Opp for the purpose of creating an implementation plan for the above-mentioned SSB/CSAA strategy that had just been published. It met in the fall of 1979 at Washington University in St. Louis.

I took special pains to be at that meeting to make sure the XTE concept was included in its report. I remember Al Opp keeping us working at writing until quite late for a couple of days. We produced a report in November 1979 entitled, *A Program in High Energy Astrophysics for the 1980s*. It described a program of large and small missions in the fields of cosmic-ray physics, gamma-ray astronomy and X-ray astronomy. It stressed the importance of the smaller Explorer missions and stated:

> The *X-ray Timing Explorer* (XTE) proposed herein will continue, with greater capability, the temporal and broad-band spectral studies carried out by OSO-8, SAS-3, and HEAO-1. Since these objectives are highly complementary to those of [the focusing] AXAF and since the XTE can be used by a large segment of the X-ray community, we urge the early initiation of this mission and, in particular, the immediate issuance of an Announcement of Opportunity.

Getting such a strong endorsement by our fellow scientists, albeit in the same general field, helped nail down the prior SSB recommendation. The date given in the report for XTE being "started" (meaning funded) was 1983. In fact, XTE was not started until Explorer funds became available for it, in 1989!

This was probably only my second time in St. Louis as an adult. I recall eating down by the river one evening and seeing the moonlit Mississippi River from the famous Gateway Arch, a breathtaking sight indeed.

XTE was not my whole life by any means; a lot was going on to keep me very busy in these years. We had just completed the MIT SAS-3 mission (1975–79) and also the HEAO-1 mission (1977–1979), and post-flight activities continued. I was heavily involved in both missions as well as with my teaching duties and my family life. My two daughters were 19 and 15 in 1979. I was also pursuing another great love, small-boat sailing in my 17-ft *Mobjack* on summer weekends. I also had had a bout with non-Hodgkins lymphoma, which was first detected in March 1976 at the height of the SAS-3 excitement. It was a stressful time, and I wonder if the stress might have been a causal factor. Successful chemotherapy and radiation treatments continued into 1977.

Gambit by a supporter; another boost by CSAA, September 1979

About this time, I was newly appointed to the Committee on Space Astronomy and Astrophysics (CSAA), for the three-year period beginning July 1, 1979. This was a subcommittee of the Space Science Board (SSB) of the National Academy that consisted of about 15 members from all disciplines of astronomy and astrophysics. (The 1979 endorsement of the XTE concept by these groups described above preceded my CSAA membership.) I was a little awed to be on such a committee and was a bit surprised to find that there was to be a review of the Explorer program at my first meeting, September 27–28, 1979.

At the meeting, in Washington, D.C., three Explorer missions were discussed: COBE and EUVE, selectees from AO 6/7 but still awaiting launch, and XTE, the newly endorsed concept, not yet selected by anybody. Fred Lamb presented the XTE case. (The week before, he had made a presentation arguing for XTE to the High Energy Panel of the 1979 Astronomy Survey Committee— more on that below.) The EUVE case was presented by Stu Bowyer, the longtime proponent of that mission.

I have always had a soft spot in my heart for Stu since he sent me a congratulatory note after our 1968 rocket flight detected X-ray pulsing from the Crab pulsar. Unfortunately, Herbert Friedman's X-ray astronomy group at the Naval Research Lab (NRL) beat us to the discovery, but by only a few days. Stu, a former member of the NRL group, wrote graciously that he only believed the NRL result when the MIT group confirmed it, which was nice to hear. (The NRL result was actually quite convincing by itself.) Our results did offer some new aspects, though. We showed that the X-ray and optical pulses were detected within one millisecond of one another, after traveling independently for some 5000 years en route to Earth from the Crab Nebula, a tied race if ever there was one.

So, this was all fine; the presentations went well. But then the discussion turned to the real issue—after the presenters had left the room—of whether or not to recommend a reversal of the order of launching, that is, to move XTE up so it would be launched before EUVE. This was a quite radical idea because EUVE was already well established in the queue for a launch, and Stu had also gone through many hoops to get it there. Any move to alter the order would be viewed as highly disruptive and highly resented. I was flabbergasted; I had no idea such a possibility was on the table. I was also highly embarrassed because of my clear conflict of interest, not to mention that I was the new kid on the block.

I was pleased that XTE was getting new attention and possibly a boost, but was this going too far? In fact, I should probably have left the room. As it was, I just sat there totally mute during the discussion. In the end, the committee chose to endorse all three missions but made no steps toward reordering the sequence. It did recommend an early issuance of an Announcement of Opportunity for XTE. This in itself was a significant step ahead for XTE.

I always wondered where the initiative for changing the ordering had come from, and I heard that Stu considered it to be me. Any warmth he might have felt for me surely went up in smoke then. However, I plead innocent! I had no idea the issue was on the table until that meeting.

It was only at the 2010 Tucson Lamb-fest meeting where I gave this talk that Fred Lamb told me it was most likely the work of David Pines, who was on the SSB at the time. He was thus in a position to suggest its sub-committee, the CSAA, reexamine the mission sequencing on scientific grounds. From a purely scientific viewpoint, this might have been a sensible question to ask. On the other hand, from a management viewpoint, it would have been highly destabilizing, because, under constant threat of rescheduling, managers of approved missions would be unable to lay out their technical and financial routes to launch with any confidence. Charlie Pellerin, then the NASA director of astrophysics, told me that, even if the CSAA had suggested a reordering, he most likely would have ignored it. I received no reply from David Pines when several years ago I asked him to comment. Sadly, he died shortly thereafter, in May 2018. Stu died in October 2020.

III
An X-ray Timing Mission (1980)

Announcement of Opportunity for "X-ray Variability," July 1980

NASA WAS THUS GETTING recommendations that it move ahead on an X-ray timing mission from a variety of sources: the Lamb-Pines workshop, its own HEAMOWG, and the National Academy. This led it to give notice that an Announcement of Opportunity for such a mission would be released shortly. A number of groups, perhaps a dozen, geared up to submit proposals for the expected Announcement. Those of us who had been pushing most aggressively for the mission had no more or less right to participate, by providing instruments, than any other scientists. It was to be an open competition, but those of us most involved in selling the concept should be able to produce highly competitive proposals.

It may have been Frank McDonald at Goddard, the head of its Lab for High Energy Astrophysics, who suggested that MIT and Goddard collaborate on a proposal. Both groups were strong in the field, and such a collaboration would probably have been unbeatable for the core instrumentation defined in the Lamb-Pines study. Thus, Steve Holt, the likely principal investigator at Goddard, and I, the leader of the MIT team, entered into intense discussions as to the nature of such a collaboration. We quickly agreed that the mission should consist of two instruments (1) a one square meter proportional counter array with small field of view, about one square degree, and (2) an all-sky monitor for monitoring X-ray source activity and detecting new sources (transients).

Finally, in July of 1980, another milestone for the XTE concept was reached! NASA released an Announcement of Opportunity for a mission addressing "X-ray Variability." Proposers were asked to address the following general scientific topics:

1. Measurement of physical parameters of compact objects
2. Physics of accretion disks, plasmas, and stellar magnetospheres
3. Geometries of source emission regions
4. Normal stars through variable mass outflow
5. Nature of bursters, transients, and Sco-X1-like sources
6. Compact extragalactic objects

But the specific scientific objectives and instrumentation were left to the proposers.

The AO further specified that the mission would fly on a Dutch satellite, TIXTE, along with the Dutch multi-pinhole (aka Dicke camera) instrument. (This instrument later flew on the Italian *BeppoSAX* in 1996 where it was known as the Wide Field Camera.) This of course again complicated the mission and applied constraints on the US portion. A second possibility in the Announcement was that the selected instruments would fly on a solely US mission. Proposers thus had to consider both possibilities in their instrument configurations.

The deadline for proposals to be submitted was October 17, 1980. The discussions between Steve and me were getting nowhere. We each agreed fully that a fifty-fifty collaboration was what we wanted, but we could not agree on the form it would take. Basically, Steve envisioned Goddard carrying the fabrication responsibilities for both the PCA and ASM, while MIT would provide the digital data processing and be the lead institution in the operations, data processing, and guest observing aspects. This was not an unreasonable stance because Goddard had experience in building large counters and had flown an ASM on the *Ariel-V* spacecraft. Steve truly felt he had reserved for Goddard only a builder's role and that he was offering MIT most of the juicy, visible, rewarding aspects. Also, by drawing on technology already demonstrated in their labs, he felt, correctly, that he was strengthening the proposal by keeping costs under control.

However, we at MIT, with our long experience as hardware builders though not in-house proportional counters, saw it differently. We (my senior colleagues at MIT/CSR and I) believed the development of an instrument was the essence of a space experiment and that doing so gave one influence over the entire mission: how it was configured and the quality of the data from it. Without any such role, we felt we would be saddled with a lot of bureaucracy while being left pretty much on the sidelines. We would be the tail on the dog, and it is not the tail that wags the dog. So, after much discussion with my colleagues and much soul searching, we decided we would compete for the entire core program, even at the risk of losing it all. We preferred to lose it rather than being at the wrong end of a phone line for many years, as we saw it. If that happened, we felt we could put our scientific energies elsewhere more productively.

I thus wrote a formal letter to Steve indicating that we would submit a proposal independently of Goddard. He told me later that he just sat there and stared at the letter, shocked, hardly believing what he was reading. My view was, and is, that his shock was rooted in his inability to see how deeply we felt about the issues we had raised.

In retrospect, we probably could have made Steve's conception work for us as it did in other collaborations. For example, our role in the HEAO-1 collaboration with the Harvard-Smithsonian group was similar to (or even less than) the role Steve envisioned for MIT and it worked extremely well, even though our formal responsibilities were minimal. I now realize that it is mostly the energy and creativity that one brings to a collaboration that determines how involved and influential one is; the formal definitions and assigned tasks count for less.

Nevertheless, our view was that formal responsibilities can matter a great deal when budgetary or technical challenges arise. We valued a primary role and believed we could do it well and were concerned that, because Goddard would also be the NASA Center managing the entire program, a collaboration with Goddard's scientists could easily become quite one-sided despite Steve's good intentions.

MIT and Goddard would thus be submitting competing proposals for the core mission. This led both groups to greater efforts to improve their proposals to the eventual benefit of the mission itself. It was clear to us that Goddard was in an advantageous position with its in-house detector capability; it had demonstrated in space both large-area and all-sky-monitoring instruments. We could offer a lot of detector experience through our usual vendor, LND, Inc. on Long Island, substantial operational experience, and highly competent science credentials, but these were not the core of the mission. We definitely felt that we were at a disadvantage. Steve may well have had similar worries, namely that NASA Headquarters might disfavor his group to avoid the appearance of conflict.

The MIT Proportional Counter Array (PCA)

For the large array detectors (PCA), we at MIT proposed sealed Beryllium-window detectors made by LND, similar to those flown on several of our previous missions, with a net effective area of 0.76 m^2 (the greater part of a square yard) and multiple-cell anticoincidence, shown to be effective in the Goddard detectors. The collecting area (7600 cm^2) was huge compared to typical detectors of the time, which were only tens or hundreds of square centimeters. It was sufficient area to collect a few X-rays in the brighter X-ray sources in a millisecond, a necessary requirement if one were to study variability on that timescale.

We then addressed the fundamental issues of how to manage the high bit rates of the PCA data stream for such sources and what kind of sky monitor was optimum. Here the creativity of my MIT colleagues really came to the fore.

The MIT Experiment Data System (EDS)

A fundamental problem for the study of millisecond variability was the limited bandwidth of the telemetry link from the satellite to the ground station. Bright X-ray sources produce lots of X-rays and each X-ray is encoded with a series of "bits," which code the pulse height (X-ray energy or color), anode number, and arrival time of the X-ray. For a bright source, with huge numbers of X-rays arriving each second, the bit rate (bits per second) required for millisecond timing would overwhelm the telemetry link by large factors if all the bits were sent.

In previous missions, the data were typically compressed to yield fewer bits per second by counting X-rays in broad intervals, or "bins," of pulse height and time. The number of X-rays detected in each of the bins would be telemetered, rather than the all the bits for each X-ray. Several of these data modes would be hardwired into the flight circuitry, perhaps one with high spectral resolution and low time resolution, and another with the inverse. The observer could select the preferred mode by command before an observation but was constrained to one of the hardwired modes. This inflexibility, we felt, was dangerous in that celestial surprises—typical in those early days—could well require a mode that had not been foreseen.

Our colleague, Garrett Jernigan, then at MIT, pointed out repeatedly that the standard binning amounted to constructing a new, secondary set of bits for each X-ray from the original time/energy bits, such that the new bits would constitute the desired binning address (energy and time). Thus, for maximum binning flexibility while in orbit, we needed a method for constructing the secondary bits according to an *arbitrary* formula during flight. That would allow us to create any time/ pulse-height binning scheme necessitated by a celestial surprise.

Garrett is an extremely creative guy who has a propensity for seeing problems in a different light than most of us. His insight was correct, but the problem was implementation. The obvious procedure was to create a reprogrammable digital computer on the spacecraft to reconfigure each event's bits to the desired binning address as the events came in. However, for the brightest sources, the required processing speed rendered this scheme prohibitive in power consumption, at least in those days—remember, this was 1980. How to proceed was a puzzle, and we even organized a small workshop of local data processing experts, hoping a solution would emerge from the discussion.

It was at that workshop that our very own John Doty came up with the "obvious" answer: table lookup. The primary bit set from a single event would be used as an address to index a table from which the appropriate binning address would be retrieved. Then, the value of the number at that binning address would be incremented by unity to indicate one more X-ray detected in that time/energy bin. The lookup would be rapid and could be done on the fly. Re-loading the table for a new binning mode could be done leisurely between observations. Why hadn't I thought of this? I knew trig functions and logarithms were often obtained from table lookups on the slow computers of those days, and perhaps still are, but its application here did not occur to me. Neither did it immediately occur to Jernigan or to Doty.

We called the logic circuitry that executed the table-lookup binning an Event Analyzer (EA). It was powered by a Digital Signal Processor chip and an additional microprocessor, which served as an EA manager. In addition to the infinitely flexible binning mode, software was written and installed for additional modes: Burst, Pulsar folding, Auto and Cross correlations, Single-Bit high speed, and Event wherein all bits of each event are telemetered. Each mode had pre-selected multiple *configurations* (e.g., bin sizes) that could be selected by an observer. Others could be uploaded to the EDS if needed.

The proposed Experiment Data System contained eight EAs so the data stream could be binned in multiple ways simultaneously, another important feature. Of course, the outputs would have to share the available telemetry. Two of the eight EAs were dedicated to the data from the All-Sky Monitor because they operated somewhat differently. Another two were devoted to two standard PCA binned modes—one timing and one spectral—that would remain unchanged throughout the flight to provide a uniform archive.

How useful was our flexible scheme in practice? We launched with a fixed set of configurations, perhaps a hundred or more, that observers were likely to need. It turned out that very few others were required postlaunch. However, the generality of the design made possible important reconfigurations of the data stream and command structure during the mission, for example to detect high-voltage arcing in either the ASM or the PCA detectors, whereby the appropriate high voltage could automatically be shut down to prevent detector damage.

This then was the essence of our proposed EDS. As finally flown (Fig. 7), it was a box about $10 \times 12 \times 30$ inches with ten circuit boards, eight of which were Event Analyzers. All in all, we considered the EDS to be one of the truly innovative features of our proposal.

The MIT All-Sky Monitor (ASM)

The design of our proposed All-Sky Monitor was grounded in several previous experiments: the scanning slat collimators on SAS-3, the single pin-hole camera on *Ariel-V*, and the multihole pseudo-imaging Wide-Field Camera being advocated by the

Figure 7: EDS with Ed Morgan and James O'Connor

Edward Morgan (left), MIT EDS instrument scientist, and James O'Connor, project technician, with the original patched-up Experiment Data System, which consisted of eight identical computers in a rectangular gold-colored aluminum case. This unit was installed in the spacecraft for the spacecraft environmental tests but, due to a failure described herein, was removed and replaced by a second nearly identical unit prior to shipment of the spacecraft to Kennedy Space Center. Ed used the unit shown here for years afterward to test software patches before they were uploaded to the spacecraft. Note the MIT logo on the side of the case.

Dutch. The goal was not only to monitor a major portion of the sky in order to detect and record activity in an X-ray source, but also to determine the position of an X-ray source on the sky with sufficient precision for the PCA, with its one-degree field of view, to reliably acquire the source by means of a spacecraft reorientation.

After much thought, we decided to adopt a camera concept that had been suggested by George Ricker of our group at MIT. It consisted of a random slit mask over a one-dimensional position-sensitive proportional counter, a one-dimensional version of the multi-pinhole Wide-Field Camera. The latter required two-dimensional imaging, a significant complication in several respects. The mask with its random slits would provide good angular resolution, to several arcminutes in one dimension. The relatively large aperture compared to the single pinhole camera being proposed by Goddard could provide much better statistics for obtaining position centroids and intensity measurements on short time scales. This, we felt, greatly offset the potential for source confusion inherent in a slit system.

As for determining source positions, such a system would yield not a spot on the sky but a line on the sky upon which the source must lie, a *line of position*, which is much longer than the PCA narrow beam-like field of view. The region of possible source positions was reduced by carrying two such cameras rotated about the view direction by a fixed angle from one another. The two lines of position on the sky for a given source would thus cross, yielding the true source position at the intersection.

We called each camera a Scanning Shadow Camera (SSC), and our proposal called for the two with crossed fields of view (Fig. 8). They were to be mounted on a single rotating platform, which would let the SSCs scan over large portions of the sky while the spacecraft was holding a fixed orientation. At this orientation, the PCA and HEXTE would be collecting X-rays from their scheduled target.

The data processing of the shadow camera data was quite involved. The software had to match the X-ray shadows created by the mask on the detectors with the mask pattern itself to obtain the celestial line of position. We initially had proposed a continuously rotating camera for which Garrett Jernigan came up with a clever binning and Fourier analysis technique, which would be quite

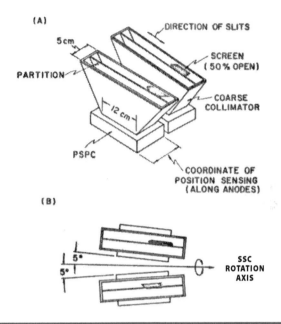

Figure 8: The ASM in the 1980 proposal

The All-Sky Monitor (ASM) in the MIT 1980 proposal. It consisted of two Scanning Shadow Cameras (SSCs) with fields of view angled 10 degrees from one another. (The angle was increased to 24 degrees and a third SSC was added on the flight instrument.) The scanning rotation axis for the two-camera system is shown in (B).

efficient given steady scanning with a perfectly stable spin axis. In the end, though, this was deemed impractical, and the scanning was carried out with sequential 90-sec exposures with the SSC orientations held fixed during each exposure.

A series of such exposures would cover about 70% of the sky as the spacecraft circled the Earth, in about 90 minutes. Spacecraft maneuvers could make other parts of the sky accessible. The ASM scanning had to be programmed to avoid viewing the Earth, which blocked the view of almost half of the sky at any given time. From the perspective of the satellite, the Earth was constantly moving across the sky, completing a complete circuit around it every 90 minutes. This too had to be taken into account.

Knowing we were up against a proposal containing a pin-hole camera, our proposal was careful to clearly describe quantitatively the relative merits of the two systems; both had strengths and weaknesses. In such comparisons, it pays to be objective and fair, because that gives credibility to one's conclusions—and I believe we were.

Proposal submissions

Our proposal, *Proportional Counter Array with a Scanning Shadow Camera for the X-ray Timing Explorer* with me as the principal investigator, was submitted to NASA on October 17, 1980. It was a great relief because we had worked terribly hard to make that proposal both innovative and accessible to the first-time reader. Our small group (Levine and Rodger Doxsey) had been joined by Jernigan, Doty, and Claude Canizares in preparing the proposal. Our conception of the instrument configuration for the dedicated US spacecraft option is shown in Fig. 9.

About a dozen proposals were submitted in all, including, as expected, one from Goddard led by Steve Holt. Rick Rothschild's proposal for the detection of high-energy X-rays (20–200 keV) to complement the core instruments was notable because it was eventually selected. Rick was at the University of California at San Diego, a member of Larry Peterson's X-ray astronomy group. He had formerly been at Goddard and was well acquainted with the large-counter technology there and knew how to complement it.

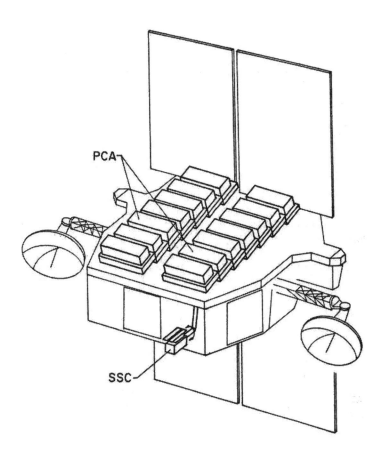

Figure 9: XTE on dedicated spacecraft; 1980 proposal

The configuration of XTE for the dedicated US spacecraft option in the MIT Proposal (October 1980). The Proportional Counter Array (PCA) consists of 14 beryllium-window detectors, each with a mechanical collimator that restricts the view to a one-degree patch of sky.

Mayer, Bradt, Levine, Schulz, Remillard, Klatt, Francis, Goeke

Top left: Bill Mayer (left), MIT RXTE project manager, and Hale Bradt, ASM principal investigator until 2001, at Bill's retirement party in 2008. Bill holds the military "gun camera" we had given him from one of the Aerobee rockets we had flown when he was a graduate student. The cameras photographed the stars and were a critical part of our data. Bill was responsible for their proper operation. Top right: Alan Levine (left), MIT XTE scientist and principal investigator of ASM after 2001 (continued page 49)

Holt, Strohmayer, Rothschild, Strohmayer, Swank

Top left: Stephen S. Holt, GSFC Scientist and principal investigator for the PCA until 1990 when he became director of Space Sciences at Goddard, ca. 1997.
Top right: Tod Strohmayer (left), GSFC RXTE scientist and NASA RXTE project scientist after 2009, and Rick Rothschild, principal investigator of the UCSD experiment, HEXTE, ca. 2005. Bottom: Tod Strohmayer and Jean Swank, GSFC PCA principal investigator and NASA RXTE project scientist until 2009, ca. 2011. [Photos: GSFC PCA team]

(continued) and Dale Schulz, Goddard project manager of the entire XTE project. Middle left: Ron Remillard, MIT research scientist, at an XTE meeting, 1993. Middle right: Brian Klatt, MIT RXTE project quality assurance manager, 2008. Bottom: Jim Francis (left), programmer of the flight software for the EDS, and Bob Goeke, MIT RXTE project engineer, 2008. [Photos: MIT RXTE team except upper right, GSFC PCA Team]

IV
Selection and Holding (1981–1989)

The George Clark missive

TO EVALUATE THE PROPOSALS, NASA formed a review committee with Bill Kraushaar of the University of Wisconsin as chairman. He had been on the MIT faculty, with tenure, but had moved to Wisconsin in the mid-1960s. He was known as an eminently fair and knowledgeable scientist; his objectivity was not at all in doubt. Fred Lamb (U. Illinois), the honoree at this workshop, was also on the committee, as were Doyle Evans (Los Alamos Science Laboratory), Terry Matilsky (Rutgers), Francesco Paresce (Space Telescope Institute), Richard Catura (Lockheed), Roger Chevalier (U. Virginia), and Frederick Seward (Harvard/Smithsonian). Alan Bunner of NASA Headquarters served as secretary.

The committee members' names were embargoed; we did

not know who they were until we appeared before them in May 1981. We were pleased that the members were all well qualified, and thus it appeared in retrospect that the proposals would be fairly judged by our fellow scientists. But there were larger issues involved in the selection that arose much earlier.

My colleagues in Bruno Rossi's cosmic-ray group had a history of space experiments in several fields (gamma ray, X-ray, and interplanetary plasma) dating back to the early 1960s. Much of it had been productive and rewarding, but of late, one or more experiences in the plasma area had disenchanted my colleagues with NASA's evolving selection procedures. It was felt they had deteriorated substantially in several respects. The evolution was probably due to ever-stricter requirements on commercial procurements, but the MIT group felt they were degrading the quality of scientific selections.

This had been the subject of strong letters between Herb Bridge (head of the plasma group at MIT and director of our Center for Space Research) and high NASA officials. One such letter written in 1977 was prompted by the categorization of an MIT plasma proposal as Category 2 by the committee. Although Herb's letter might have been written off as sour grapes, it was clear from the letter's quotes of the committee's written comments that the rating was based on an incorrect understanding. Another of my MIT colleagues, Walter Lewin, in 1978, made similar points to the NASA associate administrator about uninformed committee decisions and recommended steps to improve the process.

Moreover, in the present case (XTE), we were competing head-to-head with a NASA institution (GSFC), and NASA Headquarters would to be making the selection. Could they be objective? We were concerned that Goddard's financial well-being might just happen to carry more weight with NASA headquarters than would MIT's interests. There was a clear conflict of interest. At some later time, in discussing this problem with Carl Fichtel of Goddard, he told me the continuing concern among some Goddard scientists was that NASA Headquarters would be overly sensitive to the conflict issue and would thus be biased against Goddard proposals. This, of course, just reinforced the reality of NASA's internal conflict of interest.

All this was background noise to me as my small team was totally occupied with getting the proposal put together and submitted. It was thus a total surprise to me when our colleague George Clark—another MIT X-ray astronomer and my former thesis supervisor—revealed to us that he had sent a letter (a missive akin to a missile) to our Senator Ted Kennedy and our Representative, Tip O'Neill (then Speaker of the House), outlining our concerns with the current NASA selection process. He made no bones about the $25M the MIT proposal would bring to Cambridge—which was in O'Neill's district and Kennedy's state.

George had been supportive and helpful, but not heavily involved in our proposal effort. He had discussed his letter with none of us and probably not with our director (Herb Bridge) either. He knew such a letter would generate heated discussion about whether it was wise to incur anger at NASA Headquarters, the ultimate decider and "the hand that feeds us," at such a critical moment. He sent it on October 16, the day before proposals were due! It was George Clark's Sidewinder missile. It is attached to this document in Appendix II.

He made three points about the selection procedures, abbreviated here:

1. Proposers had no opportunity to rebut erroneous understanding or conclusions on the part of the peer committee

members, and such was increasingly likely because it was becoming common that most of the US experts in a given sub-field would be involved in submitted proposals and thus would be excluded from review committee membership.

2. The committee was only allowed to categorize proposals as to scientific worth and readiness to fly; it was not allowed to rank them. NASA Headquarters would select missions to fly from Category 1 proposals. The peer committee's judgment as to the highest priority mission for flight was thus, at least formally, not taken into account. In the recent past, the peer committee had been asked to rank the proposals, but the procedures had changed.

3. The costs of civil-service man hours was not included in Goddard proposals, thus giving a financial advantage to the government institution.

Needless to say, the Kennedy and O'Neill offices sent George's letters to NASA for comment, and they were not happily received. George had sent copies to NASA's director of astrophysics, Frank Martin, who was justly angry for not being given a chance to deal with the issues earlier. This generated an intense set of letters over the next six months that went back and forth between George and Martin and a second letter to Kennedy and O'Neill. Martin's response, simply stated, was that NASA was largely constrained by procurement law and that an Announcement of Opportunity (AO) for peer-reviewed scientific observations, as distinct from a competitively bid RFP for a *well-defined product*, was a highly vulnerable process; it could be declared illegal.[6]

In response to George's three points, Frank argued that

1. Allowing personal presentations or rebuttal, while not excluded, would hinge on the eloquence and personality of the proposer, which could introduce bias.

The obvious response: yes, but that is a lot better than bias by the eloquence of an *uninformed* prime reviewer on the peer selection committee.

> 2. NASA sometimes asks for informal rankings, but it must ultimately make the choice taking many factors into account.

Yes, but those rankings by the committee would be a valuable input to the final decision and, if formally required, could then not be ignored.

> 3. NASA makes cost allowances to level the playing field.

Fine, but will it be an open visible process, so it is clear to all that realistic numbers are used that include overhead, benefits, etc.

Selection, September 1982

NASA did modify its selection procedure somewhat in response to all this. It set up a two-phase process for the peer committee review. The committee would meet on May 21, 1981, to review the proposals. The members would generate a set of written questions for the proposers, and the next day, each of the proposers would have 30 minutes with the committee to answer them. Subsequently, the proposers would have a month to prepare written responses, and they could also submit modifications or clarifications to their proposals, within the original proposal page limits.

The committee would then reconvene to consider the written responses before making its report. Members were charged with evaluating the strengths and weaknesses of the several proposals. Then an ad hoc subcommittee of the NASA Space Science Steering Committee, composed of a group of full-time government employees (!) would categorize the proposals. The project office at Goddard would then evaluate those proposals deemed of the "greatest scientific merit" for cost, engineering, etc. The Associate Administrator for Science would then make the final selection based on these reviews.

This process gave the proposers an opportunity to counter misconceptions, but it appeared to remove the ranking even farther from the committee scientists. Nevertheless, we were grateful for the opportunity to clear up confusing points in our proposals and to make minor modifications. By then we knew that the committee had sufficient expertise to make unlikely serious errors in understanding.

The process played itself out and then, after some time (months), Steve Holt, principal investigator on the Goddard proposal, and I were informally made aware that NASA Headquarters would welcome our creating a new joint proposal containing the best elements from our two proposals. I came away from this information with the distinct impression that the Goddard PCA and the MIT ASM had both been viewed favorably, but since we had each proposed for the entire mission, a reconfiguration of our proposals was needed before those instruments could be selected.

Thus, Steve and I were again urged to collaborate on the mission. What, again? It was not exactly déjà vu because the invitation to MIT was a clear indication (at least to me) that the MIT ASM had been favorably viewed by the peer committee and by NASA Headquarters. If not, why had we been invited? Our ASM was clearly the more powerful of the two, and the Goddard group was clearly better positioned to provide a PCA given its in-house expertise.

This led to more discussion with Steve. He felt very strongly that we jointly should recommend the pin-hole camera system that Goddard could provide essentially "free."

I found it remarkable that he would still propose this when the signals seemed clear that the MIT ASM had been favored over the Goddard pinhole camera. Nevertheless, it was within our purview to select any elements that were in our original proposals to create an optimum mission.

Steve argued that the all-sky monitoring function was very important to the mission, but that it was highly vulnerable to being eliminated in one of the inevitable costing exercises we would surely encounter down the road. An essentially free, off-the-shelf, system would be much easier to defend. His argument certainly had merit, but we would not buy it.

This led to a meeting at Goddard with both of our directors present, Herb Bridge of MIT and Frank MacDonald of Goddard. We debated the merits of the two approaches inconclusively and ended with a standoff—both sides would go home and think about it. This time there was little correspondence back and forth, possibly for a month or more. I felt our case was a strong one, both politically and scientifically. When we did communicate, I simply reiterated my belief that the ASM should be the Scanning Shadow Camera system we had proposed, for the reasons stated in our proposal. Finally, sometime later, Steve conceded. I suspect his boss, Frank, convinced (or told) him that he should let MIT have their ASM.

(Here I had followed what I had learned years earlier from Herb Gursky formerly of AS&E, a friend and fellow X-ray astronomer. Over one long summer, we had many discussions as we tried to reach agreement as to which of our two instruments should fly on HEAO-1. He had the winning cards and patiently waited me out, eventually carrying the day. In the current XTE case, I felt the winning cards were held by me, so I could afford to patiently wait.)

Steve and I then worked up a memorandum of agreement that outlined a mission with the Goddard PCA, the MIT ASM, and the MIT data system (EDS) and formally notified NASA of it. It then became apparent through the grapevine that NASA was debating the possibility of including Rick Rothschild's high-energy experiment, HEXTE. It too must also have been favorably regarded by the selection committee.

It was a hard call for me because it added good science but also added cost and complexity to the mission. It did complement the PCA science nicely, and I knew that theorists valued the information in the high-energy X-rays, even though there were fewer of them relative to the lower-energy X-rays. Thus, when I received a call from Al Opp of NASA Headquarters delicately asking how I felt about having a high-energy experiment on XTE—meaning, I gathered, would I create a ruckus if NASA included HEXTE in the mission—I indicated my acquiescence. I assume Steve did also.

It was shortly thereafter that NASA formally announced the results in three separate letters from Bert Edelson, Associate Administrator, to me at MIT, Steve at GSFC, and Rick at UCSD, stating that our instruments, the ASM, the PCA, and the HEXTE respectively, had been selected. The letters were dated September 28, 1982, two full years after our proposal submissions. The letter to me stated, "It is our intent to combine the operation of your [ASM] instrument with that of the Large Area Proportional Counter proposed by Dr. Holt." This official notification was most welcome news. Another hurdle had been passed, but there were more ahead. At least now, we each were in new territory with our instruments formally selected.

One might ask if all the letter-writing about process did any good or whether MIT might have muscled its way to an unfair advantage. I doubt it had a significant effect on the selection; the conclusions were hard to fault, and the committee was quite knowledgeable. However, it could well have biased the case one way or the other because the NASA officials are humans with emotions. Was George wrong to interject such perturbations into the selection process? I think not. The issues raised were very real and a real threat to a fair evaluation. In fact, NASA's response improved the process, and the persons involved were certainly sensitized to the possible shortcomings of the aspects they could not change for legal reasons. These issues remain with us today and are good to keep in mind.

Astronomy Survey Committee endorsement, 1982

In July 1981, in the midst of the XTE proposal selections, George Clark found himself again facing the XTE issue, this time from a loftier perspective. George was chairman of the High Energy Subcommittee of the Astronomy Survey Committee at the time, and, in this role, wrote a politic letter to Hans Mark, Deputy Administrator of NASA. He described the science XTE could do and stated that his subcommittee "strongly endorsed" the X-ray Timing Explorer concept. He further stated that the Survey Committee report "will endorse XTE." The Astronomy Survey Committee was the National Academy of Sciences committee tasked with preparing the astronomy decadal study for the 1980s, so this was an endorsement of great significance. In contrast to the decadal SSB/CSAA study of 1979 described above, its purview was *all* of US astronomy, not just space astronomy.

To further the case, on August 13, 1981, the indefatigable Fred Lamb made a pitch for XTE directly to NASA Administrator James Beggs and his staff.

The Survey Committee report was published by the

National Academy Press in 1982: *Astronomy and Astrophysics for the 1980s*. The XTE concept was highlighted in the science goals section under "Study of Violent [Astrophysical] Events" and again under "The Explorer Program," where it received a complete paragraph as follows (v. 1, p. 108):

> The X-Ray Timing Explorer (XTE) will provide important new opportunities for observations of variability in X-ray sources on time scales ranging from milliseconds to years. The scientific objectives of this mission include investigations of the mass, magnetic moment, and internal structure of neutron stars and degenerate dwarfs; the physics of accretion disks, plasmas, and stellar magnetospheres; the geometry of source emission regions; the nature and evolution of normal stars through studies of mass loss; the nature of variable sources, such as X-ray bursters and transient X-ray sources; and the underlying physics and emission mechanisms in compact extragalactic objects.

XTE was now one of four Explorer missions that were in the NASA queue, but not yet launched; the others were from the AO 6/7 selection, IRAS, COBE, and EUVE. The committee gave high priority to its recommendation that NASA's Explorer budget be augmented at least to the level of the previous decade in order to expedite the implementation of these and future Explorers.

All this letter-writing, presenting, and publishing were continuing reminders to NASA officials that there was substantial support for an XTE mission. NASA had yet to give it an official "start," that is, major funding, and that was not yet a sure thing.

Waiting for funding

The spacecraft instruments had been selected. We were all set to design and fabricate, but was NASA? Well, not really. In May

1982, prior to the instrument selection, I gave a talk at the Ottawa COSPAR conference, titled "X-ray Timing Explorer." I went over the science justification and reported that the launch would be delayed until "late in the decade" because COBE and EUVE preceded XTE in the Explorer queue—IRAS was nearing its 1983 launch—and because the entire Explorer program had a very limited budget.

There then passed a period of several years during which the program was directed by the Goddard Advanced Missions Analysis Office in the person of Bill Hibbard. We needed a rocket to get us into orbit and a spacecraft to service and carry our instruments in orbit for two or more years. In NASA's planning, the Space Shuttle would be the rocket to launch XTE into orbit, despite concerns about increased costs associated with safety requirements in the manned program.

The Dutch TIXTE plan went nowhere, and we were thus left with the need for a spacecraft. Perhaps it would be the aforementioned Multi-mission Modular Spacecraft (MMS). Its proponents argued that, if NASA would buy several such spacecraft, the cost for each could be dramatically lowered. Unfortunately, NASA was prohibited from purchasing a spacecraft in the absence of an approved mission to make use of it. It was a Catch-22, but probably a good one, because each mission places its own particular requirements on its spacecraft. A spare sedan in the garage is no good to you if you really need a motorcycle or a pickup truck.

A spacecraft consists of mounting platforms for the science instruments, a command-and-control system, solar panels and batteries for electrical power, a telemetry system with antennas, and an attitude control system for reorienting the spacecraft to point the PCA field of view toward desired celestial positions. Various spacecraft possibilities were explored, but with no funds in hand, these were set aside.

We began to receive small amounts of study funds that, with HEAO-1 data analysis funds, enabled me to continue support of a single research scientist and a graduate student or two. Rodger Doxsey, who had been my student (PhD 1974) and then my postdoc on our SAS-3 and HEAO-1 programs, filled the scientist role. Unfortunately for us, he moved to the Space Telescope Institute in 1981, where he became a tower of strength in the *Hubble* project until his early death in 2009.

Alan Levine, another scientist in the MIT cosmic-ray group, had helped produce our 1980 XTE proposal. In these lean years, he picked up more of the responsibility for RXTE and was the primary ball carrier during my sabbatical leave in the winter and spring of 1983. He recalls preparing a budget for MIT's portion of the mission with Bill Mayer and being chastised at the general meeting at Goddard because the total cost was significantly higher than those of the other science teams, Goddard and UCSD. As time went on and more realism was imposed on the instrument teams, he was gratified to find that the early MIT budget turned out to be about right, whereas the others had been overly optimistic underestimates.

Every year, Bill Hibbard would assign modest amounts of funding to each of the XTE groups to support their preparations for the mission, such as building and testing a prototype detector or developing data analysis techniques. One year I had heard from him via voicemail that he was setting aside a certain amount for us, say $100K, for the next year. After some thought that same day, I called him back, but got only his answering machine, to which I explained that *another* $40K would be a great help because we could, say, take on another grad student to do such-and-such. That same afternoon, I received another voicemail from him, saying that he would increase his plan to $140K. There had been no direct voice communication, which was unusual in those days, although not these. That was the easiest and fastest $40K I ever got for our research.

Ron Remillard and Edward Morgan, both advanced graduate students at the time, began to take an interest in aspects of the XTE program. After they earned their degrees, in 1985 and 1987 respectively, and as the mission developed, the threesome, Al, Ron and Ed, became the mainstay of the MIT scientific XTE effort. How fortunate I was to have such talented scientists pulling to make the MIT effort a success. On the engineering side, Bill Mayer and Robert Goeke guided our study efforts. Bill, another former student of mine, eventually became the MIT project manager for XTE and Bob the project engineer. Bob was an MIT graduate in electrical engineering. As a freshman he had taken the basic physics course I lectured. Now as an experienced engineer, he had things to teach me.

During these lean years for US X-ray astronomy—no NASA X-ray astronomy satellites in orbit and little funding for hardware development—Ron Remillard and I sought to optically identify X-ray sources located with our HEAO-1 (A3) instrument. This took us to mountaintop optical observatories in Arizona, Chile and Australia—beautiful skies and inspiring scenery! During this hiatus in US missions we and others made good use of Japanese and European X-ray astronomy satellites (*Tenma, Ginga*, and EXOSAT) for observations. I spent a sabbatical in Japan in 1983 and visited there several other times. I also became familiar with the EXOSAT data center in Holland. It was convenient to where one of my sisters lived with her Dutch husband and children.

Theorists to the rescue again, August 1985

In the midst of these doldrums, Fred Lamb and his colleagues felt the XTE concept needed another boost. Accordingly, in August of 1985, they organized another workshop, in Taos, New Mexico. The result was a summary paper in *Los Alamos Science* entitled "Astrophysics of Time Variability in X-ray and Gamma-Ray Sources," written by R. Epstein, F. Lamb, and W. Priedhorsky. It was reproduced in a widely distributed 37-page slick brochure replete with coherent discussions of the many types of variability in X- and gamma-ray astronomy: pulsars, quasi-periodic oscillations, bursters, and potentially pulse timing fluctuations due to the internal dynamics of neutron stars. It also described the features of coming timing missions, the Japanese *Ginga,* and the American XTE.

The Epstein report gave valuable theoretical support to the science XTE would be able to do. The case was helped immeasurably, of course, by the observations of the Japanese *Hakucho* and European EXOSAT missions, in particular the X-ray burst studies of the former and the discovery and studies of quasi periodic oscillations (QPO) in low-mass binaries by the latter. It wasn't all theory!

A figure in the Los Alamos document showed the configuration of the XTE experiment payload as worked out for a Space-Shuttle launch by Goddard engineers. Notable were the two long deployable booms that held MIT's all-sky monitor units (Fig. 10). We had told the NASA engineers it was desirable to maximize sky coverage and hence to minimize blocking of the ASM fields of view by the spacecraft. This led to two long booms, each with two Scanning Shadow Cameras (SSCs) rotating about the boom axis. Given that the two booms (about 10 feet long) were at right angles to one another, essentially the entire sky would be accessible.

We were pleased to see four SSC detectors rather than the two we had proposed and were a bit surprised by the length of the booms. We were a bit concerned about the complexity of deploying them and maintaining SSC alignments. Nevertheless, we were pleased that NASA engineers seemed willing to help us optimize the science we could obtain from the mission. Beware of such help!

The following September (1986), one of the undergraduates working in our lab began research for her senior thesis, which involved developing data modes for the EDS on XTE. Her thesis was titled, *The X-ray Timing Explorer: Acquisition and Analysis of Astronomical Data*. Her later fascination with black holes may have begun during her work with us, and it has proven fruitful, because the week I wrote this, it was announced that she, Andrea Ghez, was awarded a share of the 2020 Nobel Prize in physics for her research on black holes, and in particular on the massive black hole at the center of our Galaxy. She indeed is the epitome of the previously chosen title of this work, *Chasing Black Holes*.

Challenger (1986) and Ginga (1987)

In January of 1986, the Shuttle Challenger failed on launch with the tragic loss of its entire crew. Shuttle flights ceased for the next 32 months. As the shuttle was our current default launch vehicle, this was not only a national tragedy but also a severe setback for XTE.

In February 1987, the Japanese timing X-ray mission *Ginga* was launched. Its large area of pointed proportional counters (0.4 m^2) and sky monitor were similar in concept to XTE's core instruments. The *Ginga* program surely was facilitated by the well-publicized arguments the XTE advocates had been making for years. However, we had no lock on the concept; it was the

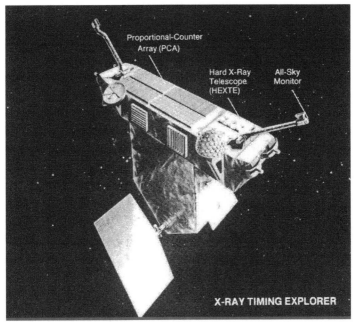

Figure 10: Goddard layout of XTE (ca. 1985)

Figure 10. XTE configured on a dedicated spacecraft for a Space Shuttle launch by Goddard engineers. The spacecraft was designed to fit in the shuttle bay. Each of the long booms carried two ASM Scanning Shadow Cameras with crossed fields of view. They were not favorably received by Charlie Pellerin of NASA Headquarters. [NASA, XTE study, about 1985]

obvious way to advance the science. The *Ginga* counters were provided by the Leicester group in England; recall that this group was to provide the detectors for our 1974 LAXTE proposal!

Ginga clearly detracted from the singular contributions that XTE might make and raised questions about the viability of a second similar mission. Ground-based optical astronomers in different countries have often built competing similar telescopes, but in the space business, because of the high costs, it is difficult to make the case for a second similar mission.

It was encouraging, though, that timing science was at least being pursued by someone. *Ginga* carried out a wealth of productive studies in both timing and spectral domains. It discovered two bright transients with its sky monitor. But *Ginga* lacked some features we were planning or hoping for XTE. It did not carry a high-energy experiment like HEXTE, its millisecond timing capability was minimal because of limited telemetry, it could not acquire new targets quickly, and its sky access at a given time of year was limited by its fixed (non-rotating) solar panels. We felt strongly that an XTE mission could make substantial advances beyond *Ginga* and hence was still worthy of flight.

Explorer platform

Our future was now largely in the hands of Charlie Pellerin, Director of Astrophysics at NASA and his deputy for high-energy astrophysics, Alan Bunner. Alan, a former X-ray astronomer from University of Wisconsin, was the trooper who kept XTE alive in the halls of NASA for many years. Charlie was a hard-driving manager whose job, with Alan, was to get astrophysics missions to flight. Success meant getting a mission to the point where it was returning data from space. We were fortunate to have both men pushing for XTE.

Charlie had to come up with a plan for getting XTE into space and keeping it there for the mandated two years. At NASA in the 1980s, the hype was all about the Space Station and the Shuttle. In January 1984, President Reagan directed NASA to build an *International Space Station*. The hype did not stop with the 1986 Challenger disaster. In late 1986, Charlie suggested that we buy into the platform-in-space and Shuttle fervor by creating a mini space station for use by both EUVE and XTE.

Sometime previously, Frank Ceppolina of Goddard had

proposed that EUVE, the astronomy Explorer mission just ahead of us in the queue, use the MMS spacecraft discussed above, and that it be launched in the Shuttle. Then, at the end of its mission, the Shuttle would retrieve and return the MMS/EUVE to Earth. The MMS could then be refurbished so XTE could use it. The MMS/XTE package would then be carried to space with the Shuttle. All this would take three Shuttle flights, but, hey, Shuttle flights were supposed to be frequent and cost effective.

Charlie modified this plan to incorporate the space-station philosophy. The EUVE and XTE would use the same MMS spacecraft, but the change-out would be carried out, not on the ground, *but in orbit* in the bay of the Shuttle! The MMS spacecraft, appropriately modified for this dual mission, would be called the Explorer Platform (EP)—a mini space station. Only one space-craft for two missions sounded like a money-saver, and it could lock the XTE mission into the planning, which might finally give it a path to space. This seemed like good news, but …

This was a bold, but possibly dangerous, concept. It required two successful Shuttle launches and entailed a complex interchange of the two payloads in the Shuttle bay, wherein XTE would mate for the *first time ever* with the EP—a standard no-no in space hardware engineering. Finally, it required that the XTE payload meet the high safety require-ments required for flight on the Shuttle because humans would be involved. This would add substantial costs to our program. Adding costs always seemed like drinking poison in small amounts. You never knew exactly when it would kill you, but eventually it surely would.

This put me, and others, in a tough moral dilemma. Should I, as a principal investigator, raise questions about this plan know-ing that in doing so, I was arguing against our most promising route into space? NASA was known for its bold initiatives, and perhaps I was just too old-fashioned to understand the potential of this scheme. On the other hand, in not questioning the plan, I could be failing in my obligations to my colleagues and to the taxpayers.

Another disadvantage to the Explorer Platform from our perspective was that the EP could in principle, host a successor experiment to XTE, and this could terminate the XTE mission as early as two years after launch, the baseline length of the XTE mission. However, with one's own spacecraft, the mission might operate for many years.

With this in mind, space astronomers had learned to design a mission "without expendables" whenever possible, meaning its operation would not require refrigerant for cryogenic cooling or a gas supply for detectors or maneuvering jets, which could be exhausted after a year or two. Astronomers would also design as much redundancy into their instruments as possible, so that most failures would not completely kill the data-taking capability. In this way, NASA might continue to support a mission financially for many years if the science yield remained sufficiently high as determined by peer reviews. But the EP could preclude this possibility. The alternative would be to formally require an extended mission. But that would require greater long-term reliability of parts and systems, which would drastically raise costs.

The EP plan gained a lot of traction at NASA, and considerable effort went into reconfiguring the MMS spacecraft for EUVE so it could take on XTE after the EUVE mission. This may well have been a factor in the subsequent delays in the EUVE launch. It certainly raised its costs.

Painful scrubbing by Charlie Pellerin, August 1987

In the summer of 1987, COBE was well on its way toward its 1989 launch, and EUVE was not far behind with what became a 1992 launch. Serious funding for XTE was becoming a real possibility. The Goddard study of XTE (Fig. 10 above) also led to a cost figure for the whole mission. As was typical, the total was deemed too large for NASA's Explorer budget. This led Charlie Pellerin to call a meeting of the XTE principal investigators at Goddard with the intent of reducing its cost. Scientists are prone to improving their experiments thus increasing cost, and it is the job of managers to keep this tendency under control. The XTE as configured did not have much beyond the core requirements, so any reduction was likely to be quite harmful. Charlie's job was to convince us that it was, nevertheless, necessary.

Well, Charlie got us in a room on August 13, 1987, and proceeded to tell us all the reasons why our mission was on the brink of cancellation. He painted a dismal picture of the state of the economy, of the public's skeptical view about NASA and its expenditures, of congressional pressures and priorities, of NASA's programmatic and budgetary woes, and the threat arising from skeptics among our colleagues who bad-mouthed XTE. In other words, XTE was no more than a fly on the wall in danger of being squashed flat at any moment, and so on and so forth. The long booms for our ASM detectors came in for particular ridicule ("like a bug's antennae") as did other elements of the payload.

Only if we could shave 20-30% off the mission cost did it have a chance to fly. (I forget the actual percentage, but it was not small.) The cuts had to be real; not just superficial reductions. Reductions in manpower without reduction in actual hardware were deemed nonresponsive; they were "only smoke and mirrors." He wanted to see genuine reductions that very day. There was some merit in his plea, because in fact, our detector numbers and areas had inflated somewhat since our original proposals; at least those for the ASM had.

It was a superb performance. No Shakespearian actor could have done it better. It was frightening and convinced us that cuts were necessary. Given the size of the demanded cuts, it was clear that the entire ASM and the HEXTE were both quite vulnerable, the latter more so than the former. However, the loss of either one would seriously damage the scientific viability of our mission, making it difficult to survive subsequent reviews.

It is very difficult to commit murder (killing another's instrument) or suicide (giving up one's own instrument) in public. Charlie suggested the scientists and Goddard managers sit down together and work through the possibilities. The managers would be able to comment on feasibility of possible cuts. However, I had learned long ago that difficult interpersonal negotiations take place more successfully in private when there is no audience.

It may have been me, then, who suggested that the three principal investigators (Jean Swank, Rick Rothschild and I, each with possibly one or two scientific associates) retire to a conference room with no NASA managers present. We were all physicists and smart enough to understand feasibility without the engineers or managers. We then went into the small Building 2 conference room, closed the door, and looked around as if to say, "Well, what now?" while our guts were turning and our brains were asking, "Will my instrument survive this?" The UCSD and MIT teams were on seemingly opposing sides; one of their instruments could well be eliminated.

Then, a remarkable thing happened, and I cannot remember who suggested or initiated it. Perhaps it was Al Levine, my MIT associate, who began to eloquently outline the value and the absolute necessity of the UCSD HEXTE instrument being on board a viable XTE. Without it, XTE would not appear substantially improved over *Ginga,* which had been successfully launched the previous February. (What? Al, that's the other team; you're running toward the *wrong* end zone!) I shouldn't have been surprised. Al had been the principal scientist working with Walter Lewin on a HEAO-1 experiment that studied the same high-energy X-rays that HEXTE would study. He could no more belittle HEXTE than he could his own family.

After we heard him out, Rick Rothschild then picked up the theme and proclaimed the many positive aspects of MIT's ASM and why it was absolutely essential to XTE. Without the ASM, XTE would be blind! This was incredible. People were not defending their own instruments, but rather the other fellow's, and the integrity of the mission as a whole. Amazing! The following discussion was very constructive and non-confrontational. It was agreed that the loss of either the ASM or the HEXTE was completely unacceptable and would damage the mission irreparably, insuring its demise in subsequent reviews.

All this back-patting wasn't getting our job done. There was no way we could leave that room with nothing to offer Charlie. Thus, each of us examined how we could cut the scope of our experiments. Each experiment took a hit. The PCA went from eight to five detectors, while retaining (or introducing) two small background detectors (PCA Jr.). The HEXTE went from six to four detectors in each of its two modules, and the ASM got rid of the two long booms and retrenched to one short rotating platform.

Rick and I both clawed back a bit of the lost capabilities. Rick made each of his remaining four detectors (per module) slightly larger so HEXTE lost only 20% of its effective area rather than 33%. The ASM, rather than cutting back to one pair of cameras on the one rotating platform, would have a third camera on the single platform. This materially increased the sky coverage. This third camera could be considered an "extra," but its cost increment over the two-camera system would be negligible, so it would be easy to defend, especially since it was a reduction from the four cameras in the engineering design.

Looking back now, it is clear that we would never have gotten that third camera unless the Goddard engineers had previously and gratuitously given us an extra two on those long booms. "NASA giveth and then NASA taketh away" (after Job 1:21), but fortunately, it taketh not as much as it giveth.

We thus emerged from the room with substantial cuts to offer Charlie, not as much as he claimed to want, but enough. He accepted them at least until they were evaluated for cost, and it appeared that XTE had survived to another day. As for the nature of the cuts and the way we managed to claw back some of the capability, he was reported to have muttered, "Those guys are awfully clever." I thought we were: we trimmed the mission substantially, but without jeopardizing its fundamentals. Best of all, we did it as a coherent team, with no bitter contentiousness.

It occurs to me now that Charlie was probably as nervous as we were about the outcome of the meeting. He had a lot riding on getting this mission accepted and in orbit; that was his job. If we had proved to be a recalcitrant group of prima-donna scientists who wouldn't pull back our goals, the mission could well have been lost. That would have been his failure as much as ours.

For our parts, we were all experimenters with substantial space experience. We had learned how to make technical and programmatic compromises in order to get our data. We had all learned the hard way that it is dangerous to push a system or an engineer too hard, because this can lead to technical and bureaucratic troubles, like being cancelled entirely, as I once saw happen to the experiment of an unrelenting principal investigator on the HEAO program. We also knew that we had to draw a line in the sand to protect the integrity of the mission so it could withstand later scientific scrutiny. Getting the right balance was, of course, the hard part.

A fateful CSAA meeting, January 1988

As the 1980s drew to a close, funding for XTE became more likely as the missions preceding it gained traction in the queue. The next step was to present the mission to the national committees of our peers. *What, again?* Yes, because it had been the better part of a decade since the 1982 Astronomy Survey Committee endorsement, and moreover the Japanese *Ginga* was now in orbit. The landscape had changed materially. Several prominent critics were still out there. I never encountered such criticism directly, but only heard about it. In perhaps two cases, though, I saw disparaging letters written to cognizant officials. More on them later.

Not being a strategist at heart, I was rather unaware of the need for XTE to regain committee endorsements. Thus, I had nothing to do with the agenda of the Committee on Space Astronomy and Astrophysics (CSAA) when it decided to review again the case for XTE. (My three-year term on the committee had expired in 1982.) Steve Holt, by now chief of sciences at Goddard, called to ask me to attend a forthcoming meeting as support for Jean Swank who would be presenting the XTE case. She was a Goddard scientist and had succeeded Steve as the principal investigator of the XTE PCA instrument.

Jean was and is a highly competent scientist who surely would do a fine job of presenting the science, but Steve knew her presentation style was rather laid back and subdued without the dramatic flair that some, though not all, value. Thus, all in all, Steve felt it a good idea to have another voice in the room, especially one with, at that time, probably more visibility in the community than Jean. I thus agreed to fly to Washington for the January meeting.

This was to be a cakewalk for me. I had nothing to present, and Jean was more than competent to do the job. This was in stark contrast to most meetings at Goddard or NASA Headquarters. At such meetings, there was almost always something big at stake—either dollars, an instrument, or sometimes even a mission. Inevitably my adrenaline would be flowing freely and my stomach churning. This time, all was calm. The committee members were academics from all branches of astronomy. The meeting promised to be pleasant and interesting.

Well, that was my mistake. After Jean's well-organized presentation, she was asked some questions, which she competently answered. Then David Helfand of Columbia University, whom I had long known and admired (and still do), turned to me—I was seated against the wall as one of the several observers—with a question I should have been ready for. He asked, quite pleasantly, "Hale, can you give us an example of a 'great discovery' that XTE could with some likelihood make?" or words to that effect.

The question was totally fair, but I was not ready for it. I had been thinking of XTE as probing the important unexplored regions of timing, not of particular landmark discoveries it might make. Great discoveries are, of course, often great because they haven't been anticipated. Yet, plausible scenarios for major progress based on current knowledge are a must to sell a mission. The Lamb studies and our proposals were full of such scenarios, but many of them were explorations of known phenomena, not unexpected breakthrough surprises. What would knock us over with astonishment? If any answer had been in my head, it might have been, "Seeing oscillations from hot spots of matter orbiting a black hole," and I might have even been right, given the later discovery by RXTE of kilohertz quasi-periodic oscillations from black hole candidates. But, without having examined that possibility more thoroughly, I felt vulnerable to accusations of theoretical naiveté by more knowledgeable committee members.

So, what was my response? It was reminiscent of Ted Kennedy's when he was asked why he wanted to be president of the US during his 1980 campaign. His and my answers can best be synthesized with "Duuuh?" though I probably muttered something general about learning about the physics of neutron stars and black holes. Jean probably tried to recoup for me, though in my intense embarrassment, I forget what transpired, but her answer did not exactly reveal the holy light either.

The committee, instead of confirming its previous endorsement, agreed that it should re-examine the viability of the XTE concept in the current time frame at its next meeting in June. I was absolutely devastated; it was probably the most embarrassing moment of my career. I took my failure as an attack on my self-image as a scientist. I was utterly depressed about how I had let down the entire X-ray astronomy community and jeopardized the entire mission.

I remember commiserating about it with Ron Remillard shortly thereafter during a long, long drive in Australia en route from Sydney to Coonabarabran for an observing run at Siding Springs Observatory. Ron was very supportive, trying to buck me up with reassurance that I had not totally negated all our XTE efforts. He was doing the driving, and I am glad this did not distract him from staying on the left side of the road, if, there had been, in fact, a center line to guide him, which was not always the case on Australian "highways" in the bush.

My failure of preparedness reminds me of another nugget that is helpful in the space business. In the mid-seventies, we had just finished presentations of the several experiments on the reconstituted HEAO program at Marshall Space Flight Center. I remember this as being quite important for the HEAO-1 program because the mission had just been reconstituted, so our presentations were a kind of defense of the newly defined mission and were thus undertaken quite seriously.

On the other hand, the HEAO-2 mission (later *Einstein*), which was also under review, was fully accepted by NASA and in place for development funding. Nevertheless, Giacconi and his collaborators gave a beautifully organized and elegant review of their mission that obviously had taken a great deal of work. I later asked Herb Gursky, a member of the *Einstein* team, why they had gone to so much trouble when the mission was already in the bag. He told me, simply, "Hale, in this business, one is always selling." I should have taken that more to heart at that CSAA meeting.

New insight at a NATO Workshop, Turkey, April 1988

In April 1988, three months after that fateful CSAA meeting, I attended a NATO Advanced Study Institute on "Timing Neutron Stars" in Çeşme, Turkey. This was right down the alley of XTE science, and I gave a talk on "Future US X-ray Programs Related to Timing Neutron Stars." This was your normal science workshop or symposium where attendees present talks on their research and listen to others' talks and watch projected slides in a darkened room. There was time between sessions, though, to admire the waters of the Aegean Sea and to visit the rug shops downtown.

Let me digress to point up the value of such conferences,

lest they appear to be boondoggles to the reader. First, they are located in various countries, including of course the USA, because science is an international enterprise. To attract attendees, they will often be located in or near major travel hubs or at attractive resorts, though in the off-season to minimize costs.

In preparing for such conferences, one is always stressed about getting one's paper ready for presentation; it's like an exam in school. This often means a rush to complete some aspect of one's research, which gives a forward impulse to the field. There is nervous pressure before, during, and after, when the talk has to be written up for the proceedings. A conference creates the time and place for informal discussions among scientists. Ideas are traded and argued, collaborations formed, and papers are begun over coffee. This too pushes the science along and helps the taxpayer get maximum science value for the dollar. While they get scientists to nice places, it is definitely not just lounging around the pool. But there is some of that too, and some of those conversations do take place in the pool.

As an illustration of such progress, I recall three conversations with colleagues at this Çeşme meeting that related to XTE.

(1) Fred Lamb patiently explained to me, using "loose screw" as a metaphor, how spin-rate fluctuations in neutron stars could probe their interior structure. He explained, "If you kick something, it will respond differently if there is a screw loose in it." Now this might seem simple to the lay person, "Who wouldn't understand that?" I could see that the "kick" could be the torques being applied by accreting matter, and how timing pulses from a spinning star could reveal spin-rate jitter. But I had trouble seeing how such data would actually yield useful structure information. I remember his explanations as we sat on a pile of rugs drinking tea in one of the Çeşme rug shops.

(2) Also present at the conference was the aforementioned David Helfand, a member of the CSAA, a well-regarded astrophysicist who was, so I had been told, a vocal critic of XTE. I had also heard that he had been particularly irritated at some comments I had made publicly at some meeting— perhaps pertaining to the small amount of XTE funding we received during the study phases. I do not recall making any such comments, but it could have happened, and it could have seemed arrogant to those with no such funding.

I had long known and liked Dave, and he was always respectful and friendly to me. I had sensed no personal antipathy on his part, so perhaps the rumors were false, nor had I personally ever heard reservations about XTE from him. David was, and probably still is, quite an independent-minded person, not afraid to buck the establishment, as when he refused as a matter of principle to accept a tenure appointment at his home institution, Columbia University. He preferred to be judged on his performance at regular intervals, and by extension felt that others should be also.

With the new CSAA review scheduled a mere two months later, I asked Dave to sit down with me for coffee to talk about XTE, one on one. I began by saying that I considered my efforts on behalf of XTE to be more of a service to the community than of benefit to me. I also pointed out that XTE was nearing reality in NASA's planning and that killing it because one might feel it not worth the $200M cost would not release the money to another astronomy project. It would be a lost opportunity for astronomy. I asked him to look at it from the community viewpoint and to set aside any personal feelings he might have about me. All this was received graciously and with no real debate about the actual merits. I left the conversation with the warm feeling that we had reconnected and that he would view the mission fairly, if not favorably.

(3) At this time, I was still reeling from the CSAA debacle and had been mulling long and hard on the most cogent arguments for advocating XTE. It was at Çeşme, in the line at a cafeteria-style restaurant—with all those wonderful Turkish foods—that my friend, Jan van Paradijs, now sadly deceased, made a great point. He simply mentioned to me that the *dynamical time scale* for matter near neutron stars was in the millisecond range, and that that should be the fundamental argument for XTE because it was designed to study such time scales. This time scale is the "free fall time" for matter in the presence of gravitational matter. It is proportional to the inverse square root of the matter density. The greater the density, the shorter the time scale.

Now I knew that millisecond time scales were important for neutron stars and stellar black holes; for example, orbital times in the near vicinity of neutron stars were about 1 millisecond, and I knew about the dynamical time scale. However, the general applicability of the dynamic time scale to XTE's objectives had not occurred to me. It says that *any* gaseous motions in the near vicinity of a neutron star or stellar black hole *must* occur on a time scale of milliseconds. XTE would be observing such motions because X-rays originate directly from the extremely hot gases in the vicinity of such stars. The dynamical time scale is a more fundamental and general concept than the more specific orbital time scale or free fall time. It also sounds more profound!

From then on, if there was some erudite conversation about the science that XTE might or might not be able to do, possibly so erudite that I could barely follow it, I could interject in an offhand way the comment that "and, of course, the *dynamical time scale* of matter near a neutron star or stellar black hole is right in the range XTE can study." I used that successfully in more than one critical committee meeting. You could almost feel everyone pause and think: "Oh, yeah; that's so, isn't it." Thank you, Jan, for that.

I can hardly remember some of the places I gave talks in those years. However, I will never forget Çeşme. Nor will I cease thanking my friend, Hakki Ogelman, also now deceased, for organizing it.

Critical letters

In the interim before the forthcoming CSAA meeting, two letters strongly critical of XTE became known to me. They illustrated the hurdles facing us. In one case, the arguments were measured and sensibly presented, but the other was absolutely virulent. It was most impolitic and dismissed XTE as severely and irredeemably flawed in its science goals, its method of selection, and cost. I have no copy of the latter and only saw a retyped version with no indication of the author, though rumor associated a name with it, the name of a civil servant highly visible in the NASA system. That possibility was so shocking that I still entertain the possibility that the letter was fraudulent or misattributed.

The more sensible letter was written by David Helfand, within weeks of our Çeşme conversation. So much for my political skills! I had been sent a copy but had completely forgotten it until reminded of it recently, in 2013, by Fred Lamb, who was able to find a copy; see Appendix II. It was an honest and sober assessment of XTE from David's perspective, with arguments that deserved serious attention. The arguments in both letters were similar: old technology, a flawed selection process, science objectives that were being preempted by *Ginga* and other missions, and of course the high cost.

It was the newness of the science, not the technology, that mattered. Yes, *Ginga* cut into some of XTE's objectives, but XTE had a lot more to offer—effective millisecond timing, high-energy sensitivity, a powerful all-sky monitor, and, we hoped, the ability to point to any sky position (except close to the sun) rapidly and at all times. All that opened up a lot of discovery space.

The flawed-selection criticism had some merit because, as stated, there had been no formal Announcement of Opportunity for an Explorer mission from which X-ray Variability emerged as an approved mission concept, though it had appeared in tentative selections. That and many studies and endorsements led NASA to issue an AO for a mission addressing generally, "X-ray Variability." How narrowly or broadly an AO should define the goals of a mission is an arguable point; recall the AO 6/7 experience. The bottom line for me is that the politics for bringing an expensive scientfic project to fruition is so complex and ever changing that there is no sure path to success. Bravo to those few who manage to traverse and survive the labyrinth successfully and ethically.

Finally, we felt that David's personal tastes affected his conclusions. Not all possible scientific investigations are worthy of pursuit, but among worthy ones, scientists can differ on their relative importance. Variability studies of neutron stars and black holes were high on our list but apparently not so high on David's. Nevertheless, the challenge of meeting these arguments really focused our minds and materially improved our presentation at the forthcoming CSAA meeting.

A 100% Guest Program

In the midst of all this, Steve Holt came up with a winning idea, namely that every bit of XTE data should be made available to the general guest-observer program. The instrument teams would reserve no data for themselves and would compete for observing time along with all others.

In the early days of X-ray astronomy, all the data from an experiment belonged to the scientists who proposed and built the experiment. This was in the classical tradition of physics experiments. However, the morality of this soon became suspect because these experiments were built with public (NASA) funds, and such funds also paid for the launch vehicle and the data analysis efforts. Maintaining public support for such missions argued against a single investigator or institution reserving (hogging) all the data. Why should investigators in Massachusetts have all the data that was paid for in part by citizens in Michigan when there were highly competent scientists there who could make good use of it?

"Guest" programs on space missions began in a small way in the 1970s with outside investigators collaborating on observations with the instrument scientists. But then NASA began to set aside small amounts of funds and observing time for guest observers, and then began to require that proposers of missions include guest programs in their proposals. The size of those programs grew to become the dominant portions of the observing programs, but always the instrument principal investigators retained a significant part of the observing time, especially during the first year or two. Guest programs required extra work, time and money, because the hardware systems had to be well-documented and the data analysis programs transparent and usable by guest scientists.

In addition, it had become standard practice that data were reserved for an observer—whether an instrument scientist or a guest—for only a limited time, typically one year, after which the data became available to all. Gone were the days when instrument principal investigators could fuss with the idiosyncrasies of the data and take their time in publishing results.

Steve's proposal was unique in that we, the principal investigators and our teams, would retain *no* exclusive rights to any portion of the data at any time. What about all the creativity, sweat and tears we had put into our experiments, from conception to launch? Didn't we deserve a payoff of having at least some portion of the observing time guaranteed to us? Yes, but didn't we also want to sell the mission? Hadn't the whole exciting experience of developing an instrument for flight with NASA funds been rewarding in its own right in that it had brought additional young scientists into our organizations enlivening our intellectual environments? And finally, Steve's proposal did not shut out the instrument teams. Their individual members would be able to propose competitively for observing time along with scientists worldwide. Since they knew the instruments better than most, they would be in a strong position competitively.

After receiving Steve's call with his proposal, I mulled it over awhile and discussed it with my younger colleagues. What was fair to them and to the community. What was right? Without much difficulty, we came down on the side of "100% Guest." Ron Remillard in particular was always sensitive to, and vocal about, the viewpoints and rights of other scientists, whereas my tendency was to be more concerned about the rights of my associates. However, even I was quick to see that 100% guest participation would be beneficial for XTE in many respects.

It was also possible, we realized, that a competent individual on an instrument team could well obtain *more* observing time than he or she might with guaranteed time. In the standard model, after the first year or two, no more than about 25% of the total observing time—the time instruments are actively viewing the sky—would be reserved for the instrument-team observers. But, dividing that among perhaps 15 instrument-team scientists at our three institutions could be quite limiting. With the opportunity to compete freely for observing time, there would be no such limit.

The other instrument teams agreed with this approach, and this plan would be clearly enunciated in the forthcoming report to the CSAA. We reserved only the first 30 days for calibration of the instruments, and the data from those observations would be released immediately to the community. All of the observing time thereafter would be allocated by a peer committee based on competitive proposals, which would be judged on their scientific merits.

The ASM data required rapid analysis for alerting the community to new X-ray sources or interesting changes in known sources. It would be like a continuous weather update and therefore not particularly amenable to observing proposals. My group thus planned to carry out rapid analyses to obtain source intensities and to make the light curves—plots of intensity as a function of time—publicly available in near-real time. These light curves would show the behavior of the approximately brightest 100 X-ray sources in the sky. They would be available in digital form to all to use for further analysis, for planning observations with XTE or other observatories, or for showing context for a completed observation. This did not exclude guest observers requesting and working with the raw ASM data, but that right was rarely exercised.

Report to the CSAA, June 1988

The next CSAA meeting in June 1988 was fast approaching, and we needed to produce a convincing in-depth report of the science XTE could address. Without current CSAA endorsement, the mission would most likely be cancelled. The comprehensive documents from the Los Alamos and Taos workshops were old hat by now, dating from 1979 and 1985, both pre-*Ginga*.

In preparation for that report, Steve Holt proposed that we go to the community for letters of support, which we did. On March 14, we wrote 45 of our colleagues in astrophysics worldwide asking them to write how they would use XTE to further their own science interests. We received 39 responses with what amounted to brief science proposals for observations. They were for studies of a variety of objects: stars of all types, black holes, supernova remnants, and active galactic nuclei.

The letters showed the breadth of interest and the richness of the science in a concrete way; no smoke and mirrors there. The writers included many prominent scientists: M. Burbidge, M. Elvis, E. Feigelson, R. Giacconi, D. Lamb, F. Lamb, A. Parmar, H. Ogelman, N. Shibazaki, I. Tuohy, M. Urry, J. van Paradijs, and M. Weisskopf, among others. Most were from the US, but Europeans, Australians, and Japanese were represented. Together they made a powerful impression. These letters would be a part of the report to the CSAA. Steve sure earned his pay the day he suggested this.

The body of the report came together with the writing help of experts in the several science areas, and it emphasized the actual observations that would probe defined questions. I believe Jean Swank, as project scientist, took it upon herself to organize the scientific contributions, as she did so many times for peer reviews after launch. The most important section of the completed report was headed: "The Key Scientific Objectives of XTE," which consisted of seven objectives such as the "Structure of Accreting Neutron Stars" and "Behavior of Matter Close to Black Holes."

Each of these was accompanied by one to three scientific topics, for a total of 15, e.g., "Angular Acceleration Fluctuations"—that's the loose screw topic mentioned above—and "Neutron-star Spin Frequency," each with specific observations XTE could carry out. There were also lesser sections on "Multiwavelength Science," "X-ray missions of the Next Decade," and "XTE Community of Users," which included our 100% Guest Program.

The report, *A Study of the Scientific Objectives and Capabilities of the X-Ray Timing Explorer,* thus far amounted to 66 pages, quite a bit for a committee member to read in its entirety. To make the report more accessible, we added a much more compact "Summary Report," which amounted to 14 pages. Here we synthesized each of the eight objectives (we divided the seventh) and pulled out and specifically labeled select observations, of which there were twelve. The 14 pages were more than we had hoped, so we preceded it with a "Synopsis" of three pages. We thus had a version for everyone, except for those who wanted a one paragraph abstract, which we did not provide. The synopsis emphasized *very briefly* the actual observations XTE would do. For example, I present here *in its entirety* the text under "Behavior of matter close to stellar black holes":

> XTE will determine the character of millisecond variability (aperiodic and quasi-periodic), which will provide strong diagnostics of the innermost regions of accreting black holes, including possibly (quasi-) periodicities from relativistic matter in the innermost stable orbits.

Among our team, preparation of the report generated a lot of enthusiasm and new confidence in the merits of the mission. I remember the final days of preparation working on it at Goddard and believe the summary and synopsis were largely my creations. There was some chuckling about how short a synopsis had to be for a typical committee member to fully absorb. The entire bound report with peripheral materials came to 92 pages. In addition, we attached the 39 letters with the observation proposals. These letters were not circulated with the more widely distributed copies because some of the ideas in them were proprietary. The document made a powerful, convincing case for the mission. We were all quite proud of it.

At one point, Steve came by while we were discussing how the presentations would go. Steve suggested that the report should not be passed out to the Committee until *after* the verbal presentation because, otherwise, the committee members would busy themselves perusing it and thus not focus on the presentation. We did this for the HEAMOWG and possibly for the subsequent CSAA presentations.

We had been planning that I would present the case to the HEAMOWG and Steve the case to the CSAA. The former would be a kind of rehearsal for the CSAA presentation. The HEAMOWG, consisting of high-energy (X-ray, gamma-ray, and cosmic-ray) astronomers, was presumably an easier sell. The CSAA was the big deal, covering all of space astronomy. Since I had worked hard on the document and was fully familiar with the science and observations described therein, I was raring to go before the HEAMOWG. I also thought it would be efficient for me to go before the CSAA with the same pitch, instead of Steve who would have to familiarize himself with some of the details.

I had the temerity to suggest this to Steve in the hallway in Building 2 at Goddard. He became really upset and vocal, and I responded similarly. When I saw our emotions ratcheting up, I suggested that we continue the discussion in private, knowing that privacy can be calming, and pointed to an adjacent empty office. As we made steps toward the office, Jean Swank who was standing nearby moved bodily between us, calling out, "No, no, no!" Steve and I had to assure her that this was not a western bar-room fight heading outside for fisticuffs or worse; we just needed privacy to calm ourselves down.

Once inside the room, Steve explained to me that Charlie Pellerin had asked him to give the presentation and there was no way he could back out of that. I, of course, readily withdrew my suggestion, and, in a matter of a few minutes, all was again sweetness and light between us. Steve was nice enough not to mention my shortcomings at the previous CSAA meeting as a factor, and maybe it wasn't. But then again that may have been why Pellerin wanted Steve to do it. After all, Pellerin had a lot riding on this too.

My presentation to the HEAMOWG, on June 2 or 3, 1988, went quite well I thought. Given the beautiful and organized case made in the report, it was not difficult, as my talk tracked the summary of the report closely. One member, in the question period, commented that the presentation was so effective that Bradt should make the presentation to the CSAA also. I appreciated the comment but countered that that was not an option. I was not going there again!

Our report cover showed a sketch of the XTE on the Explorer Platform (Fig. 11) as envisioned by me with no help other than Apple's MacDraw program. The ASM was restricted

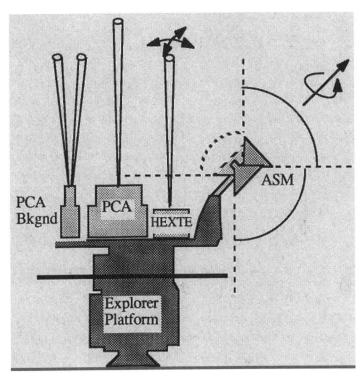

Figure 11: Imagined XTE layout for CSAA report 1988

Schematic of XTE on the Explorer Platform as envisioned and drawn by me, after the Pellerin de-scoping, for the cover of our June 1988 report to CSAA. (The actual configuration in orbit is shown in Fig. 2.) The ASM has three scanning shadow cameras (SSCs) on a rotating platform, one directed along the rotation axis and two viewing azimuthally with fields of view rotated 24° from one another. I used this sketch for the COSPAR XXVII meeting in Espoo, Finland, July 1988 [Adv. Sp. Res. 10, 297 (1990)]. That talk was given for me by Claude Canizares because I was getting my coronary arteries expanded at the time.

to one short rotating mount and, though not apparent, the PCA and HEXTE had the reduced areas that resulted from the Pellerin shakedown. Notable was that there were small detectors with offset fields of view from the PCA for measuring background. We called them "PCA, Jr." They were removed in a later cost-reduction exercise.

I was not at the CSAA presentation, which took place on June 23. Steve gave an overview that briefly summarized our report. This was supplemented by short 15-minute talks by three members of the wider scientific community: Fred Lamb (who talked on neutron stars), Jeff McClintock of Smithsonian/ Harvard (black holes), and Richard Mushotzky of Goddard (active galactic nuclei). All three had strong research interests that would benefit from XTE, but each could claim some independence from the groups that were implementing the mission and receiving funds from it. The committee then gave XTE a strong endorsement. This gave NASA the assurance it needed to start the XTE program, that is, to release the money for the design phase.

All in all, the report to the CSAA was a hugely beneficial exercise. It crystallized the case for XTE in the current era and got a large team of scientists thinking about the science they could do with it. Perhaps my blunder at the January CSAA meeting served a good purpose in the end.

With the CSAA endorsement, there now appeared to be no impediment to NASA getting the instrument teams under contract and for serious design and building to commence. But there were still hurdles ahead. Contracts required negotiation between NASA and the universities (MIT and UCSD), and the spacecraft issue was not yet completely settled. And there was still another cost scrubbing ahead. Would it never end? Nevertheless, we felt we were now on a solid path to funding, implementation, and flight.

V
Fabrication Adventures (1989–1995)

A problematic Technical Direction Clause

THE XTE INSTRUMENT TEAMS WERE brought under contract in 1989. Each institution had to provide detailed budgets for design and construction of their instrument(s) and for two years of post-launch activity. That in itself was a lot of work. We were fortunate that Bill Mayer would be our (MIT) project manager with Bob Goeke, electronic engineer, as his deputy. These two had long experience with managing large space projects and knew the intricacies of planning complex jobs with financial spreadsheets and PERT charts, which kept track of all the tasks involved. Most important, they knew how to deal with NASA's engineers, managers, and bureaucrats. They were backed up by staff at our Center for Space Research, which is where my group worked. The XTE engineering and technical staff were housed in rented space in the Albert Hill building (NE80), about a five-minute walk from our campus offices.

The MIT contract for the implementation, launch support, and two years of postlaunch activities (mostly science) came to $25M, a huge amount from my perspective. Of course, it would be made available to us only gradually over the years as it was needed. It would support several scientists and a fairly large engineering and technician team. The former included Ron, Al, and Ed, a graduate student or two, and my summer salary. The engineering team would design, construct, and test the ASM and EDS. The scientists would work closely with them during the design and testing of the instruments to help ensure their correct operation. The contract would also cover purchases of materials and detectors from outside vendors.

My scientists had an eye and appreciation for good engineering, and Bill and Bob had an appreciation for the science we were trying to accomplish. Bill and Bob and their entire team were in a separate organization within MIT's Center for Space Research, the Laboratory for Space Experiments (LSE). Bill reported directly to the head of LSE who reported to the CSR director, not to me, the principal investigator. MIT could better track progress and control costs this way. Also, it freed the scientists from bureaucratic issues and procedures, at least in part. And it helped protect the management structure from undue scientific interference in the fulfillment of contractual obligations.

Nevertheless, for the most part, we scientists and Bill's team worked closely together. I recall no divisive confrontations. Despite having been my student, Bill was in no way beholden to me. He was his own man and always seemed sure-footed in his view of the path to follow. I usually saw him as my senior in his broad view of the managerial and technical issues facing us. As manager, he ran the show. Occasionally Bill would bring me into play when, on occasion, he needed clout to make a point with NASA or the MIT administration.

For example, as we were negotiating the contract between MIT and NASA/GSFC, Bill approached me to explain a problem. There was a clause in the NASA contract that we at MIT did not want included. It was known as the Technical Direction Clause. It stated that NASA could direct MIT on how to proceed if a technical problem arose. This made sense if NASA were purchasing a device for the spacecraft, such as a power supply, but it subverted the essence of a scientific investigation. The scientists carrying out an investigation were the best judges of how to get around a problem while maximizing the science yield.

Direction by NASA would more likely be guided mostly by programmatic issues (cost, schedule, and functionality) to the detriment of the science. Of course, the scientists' solution had to comply with budgetary and schedule constraints, because NASA could refuse to provide added funds. The Goddard project office claimed that, of course, they would factor in the science goals, but Bill assured me that the Direction Clause could lead to real harm down the road. He needed visible principal-investigator support from me to get this removed from the contract. I agreed and weighed in with authority despite my superficial knowledge of such matters, and we did succeed in getting the clause removed.

Bill recently gave me more background on this:

The NASA Federal Acquisition regulations at the time exempted universities from the Technical Direction Clause, so we were surprised to find it in the proposed XTE contract. When we pointed out to the Goddard contracting officer that MIT is a university (and hence exempt), he admitted that he had thought it important to include the clause and had obtained a waiver from NASA Headquarters to include it, without notifying MIT or UCSD. It now would be embarrassing, he said, to go back and ask Headquarters to un-do the waiver. We tried to negotiate a middle ground, but he wouldn't budge. We got the MIT contracts office on our side because they didn't want a precedent to be set, and they sent a letter to him saying MIT wouldn't sign a contract with the clause included. UCSD did the same thing. In the end, he had to go back to Headquarters and get permission to remove the clause. He was replaced shortly thereafter.

The project team at Goddard took pride in their ability to hear and act on our concerns, but we knew that, sometimes, bureaucratic pressures could override good intentions. We, being the managed rather than the manager, felt we needed formal protections for the occasional crunch. Indeed, a few such issues did arise during implementation that caused us to be thankful the Technical Direction Clause had been removed.

A dedicated-spacecraft study and another painful scrubbing, April 1990

As the instrument teams began their design efforts, in 1990, the XTE was still to fly on the Explorer Platform with a change-out of instruments (EUVE to XTE) to take place in the Shuttle bay. The risks associated with this complex operation were becoming apparent upon closer examination. Also, the 1986 loss of the Challenger and subsequent 2.7-year hiatus of Shuttle flights made clear that our dependence on an operational Shuttle carried substantial risk of additional large delays.

An alternative plan was thus revived, namely the fabrication of a dedicated spacecraft designed for XTE, which would be launched by a Delta II rocket. This would completely free the XTE project from the Shuttle, the Explorer Platform, the EUVE project, and the (expensive) manned space program. Goddard would design and build the spacecraft and integrate the experiments into it, making use of its own expert in-house team of engineers and technicians.

A study by Goddard of this plan yielded a cost for the spacecraft. In the same time frame, new costs were generated by the three instrument teams. The total program then underwent a review led by George Newton of NASA Headquarters. As usual, the total cost had come to a higher value than was deemed acceptable, and there was again great pressure on both the spacecraft people and the instrument teams to bring that cost down. If we were to escape from the risky Explorer Platform, it was essential that the dedicated spacecraft plan be made attractive to NASA Headquarters. Again, the mission was at a critical juncture!

I remember sitting around a big table, possibly in April 1990 at Goddard, with Dr. Newton and his staff from NASA Headquarters, managers and engineers from the Goddard spacecraft group, and my fellow scientists from UCSD and Goddard. From my team, there was Al Levine, Bill Mayer, and Ron Remillard who was new to such meetings. Everyone was put on the spot to give up something to get the cost down. It was made clear again that smoke-and-mirrors savings such as reductions in man-hour estimates were not acceptable. Only giving up real hardware or complete tasks would count.

At stake were the integrity of the science experiments and the spacecraft systems that we desired for an effective mission. The latter included rapid maneuverability, nearly continuous telemetry at a high rate, and (nearly) all-sky all-year pointing of the PCA. If the ASM saw a spectacular flareup of a source, we wanted to be able to reorient XTE to bring it into the PCA narrow field of view and collect data from it, no matter the time of year or where the source was on the sky.

This required rotatable (not fixed) solar panels for electrical power and two antennas, one of which would be kept continuously pointed toward one of NASA's two TDRSS telemetry relay satellites. The two antennas would allow XTE to download data and to upload commands from almost any point in the XTE orbit regardless of XTE's orientation. This was an important attribute of our mission.

Consider XTE to be like a diving champion in mid-flight doing flips on the descent. As XTE moved extremely fast in its orbit around the Earth, it could reorient itself ("maneuver") about its center of mass, just as a diver reorients (flips to a new position) while falling. In this manner, the narrow fields of view of the PCA and HEXTE could be put on a celestial target of interest.

At the meeting, the pressure on us was huge to give up one of the two antennas and the transmitter that went with it. It was clear, though, that the Goddard spacecraft engineers did not want to lose it either. They considered the two-antenna system to be elegant and important for simplicity of spacecraft operations. With it, one could always download data and upload commands without reorienting the spacecraft; reorientations could severely handicap the science observations. Nevertheless, the headquarters people kept the pressure on, hinting that the alternative might be to remove an experiment.

We took a break for lunch, and Rick Rothschild (principal investigator of HEXTE) was behind me in the lunch line. He leaned over and quietly said, "Hale, I think we are going to have to give up that antenna and transmitter." Rick knew that without that concession, his experiment could well be eliminated.

My response was nearly (or perhaps I really did it) to grab him by the shirt collar with my face inches from his, to vehemently say, "Rick, don't give up now; that antenna is terribly important, and even the Goddard engineers are in favor of it!" With his stiffened backbone, the second antenna-transmitter system survived. Rick recalls the rescued transmitter as a high point of those negotiations. We were glad we had two because one of them failed well into the flight some years later.

Unfortunately, other reductions of the spacecraft systems did not reach the cost level desired by George Newton. The rest was up to the scientists. Again, we scientists went off to a private room to see what we could come up with. From the previous de-scoping, the instruments were at the edge of scientific viability. So, again, we were on the spot. With some effort, we did manage to find some meat for the lion.

Jean Swank gave up the PCA Jr. background detectors because her group had come to the conclusion that background could properly be determined without them. Rick gave up building a prototype version of his experiment, something he considers to this day a mistake. It would have revealed a dead-time problem in his data logic that later had to be tolerated in flight.

MIT gave up the purchase of additional spare detectors for the ASM. This was a highly dangerous move because the detectors were technically complicated, and spares could be essential for resolving future technical issues. However, we had nothing else to offer the cost-scrubbing gods that was real and not smoke and mirrors. Ron Remillard remembers being appalled at the loss of the spare detectors and quietly asking Bill in the meeting how we could possibly get by without them. Bill responded quietly to him, "Don't worry, you'll get your detectors." Thank goodness we did, because we needed every last one.

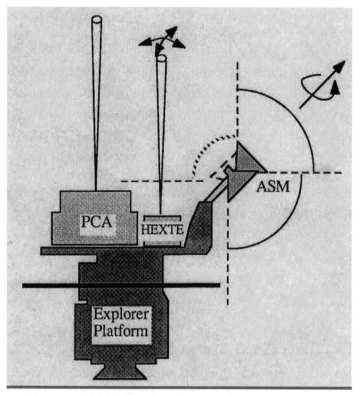

Figure 12: Imagined XTE layout after Newton review, 1990
Imagined instrument complement on the Explorer Platform after the de-scoping by George Newton in 1990. Note the missing PCA background detectors that are seen in Fig. 11. This was an easy design modification to make because this entire configuration was totally my creation with help from MacDraw.

Fortunately, these offerings were enough. All in all, we survived this scrubbing, but we were sure getting thin.

The result of the Newton meeting was a viable plan for XTE on a dedicated spacecraft, but this plan had to be proposed to and accepted by NASA Headquarters. The basic plan was still to use the Explorer Platform with change-out from EUVE to XTE in the

Shuttle bay. My vision of the XTE on the Explorer Platform (Fig. 12), required only one modification due to the Newton review. The PCA Jr. detectors (Fig. 11) were removed.

A dedicated spacecraft would be ideally matched to our goals. As conceived, and as finally realized (Fig. 2), it was quite autonomous. To reorient it, one needed only to command it to the new orientation. Onboard software would take care of solar panel rotation, antenna pointing commands, and also the roll-yaw-pitch maneuvers, while avoiding PCA exposure to the sun. And most important, thanks to rotatable solar panels and pointable antennas, it could reorient itself to put the one-degree PCA field of view on an X-ray source almost anywhere in the sky at any time without losing its power source, the sun, or breaking the telemetry link. It was the ultimate dream machine. But as yet, it was still only a dream.

The spacecraft would be built and integrated with the instruments right at Goddard, which would obviate a search for a spacecraft vendor. The downside was that building a new spacecraft for XTE would probably raise the price as compared to reusing the Explorer Platform. But the Shuttle costs associated with the launch and change-out weighed heavily on the other side. We hoped and prayed that NASA would adopt the dedicated spacecraft for XTE.

Budgeting stratagems

It turned out that we were able to order the spare ASM detectors while absorbing elsewhere the decrease in funding. We could do this thanks to Bill Mayer's costing strategy. There are two kinds of reserves in space instrument budgets: those that are open and visible and required by NASA, and those that are not readily apparent. In our case, the latter was a conservative over-estimate of the required technician labor. This gave us the financial leeway to make the spare purchases. The Goddard managers were not aware that we purchased them because, while they reviewed our financials monthly, they did not see our purchase orders because we were an independent entity (a university).

This lack of forthrightness engendered by the system is disturbing, but how else could the system work? The pressures on management to hold down costs is severe but so is the pressure on the instrument builders to preserve scientific capabilities. The fudging of the truth is especially insidious when organizations underestimate the cost of a project to win a contract. In our case, the fudging was in *overestimating* the true cost of a *trouble-free* development, so we could manage the unexpected problems that always arise in such first-time projects. This is sometimes called, "planning for the unknown unknowns."

And indeed, we did deliver the instruments to NASA on budget, i.e., within the previously defined Total Estimated Cost (TEC). Had we argued in the Newton review that we could lower cost by simply eliminating some technician labor in our budget (the truth), it would have been dismissed as smoke and mirrors. In fact, while executing the program, Bill kept the project office at Goddard informed verbally of the true status of our expected runout cost.

Bill's planning also made possible the solution of a major technical problem with the EDS, described below, without exceeding the Total Estimated Cost. However, there was an unexpected cost savings, which helped us do so. At the time of the XTE launch, another instrument program at MIT (ACIS for *Chandra)* was ramping up, and staff who had been budgeted to help with XTE postlaunch issues were needed for that program. When XTE launched, they moved to the ACIS project, which then carried most of their salaries. Since they were still at MIT, they were available for consultation at minimal cost to the XTE program.

Explorer Platform change-out revealed

In July 1990, I gave a paper at a COSPAR meeting in The Hague wherein I showed NASA engineering drawings illustrating the change-out procedures if we were to use the Explorer Platform. The procedure (Figs. 13–15) may be summarized as follows: EUVE would be carried into orbit mounted on the EP by the Space Shuttle. The EP/EUVE would be released into orbit and would carry out observations for the scheduled one or two years. When the EUVE mission was complete and XTE ready, the XTE instruments would be carried into orbit in the Shuttle as a single package called the XTE Payload Module (PM=XTE) attached to a Flight Support System (FSS), which fits into the cylindrical Shuttle Bay.

Once in orbit, the Shuttle would rendezvous with the EP/EUVE and capture it with the Shuttle's Remote Manipulator Arm (System) (RMS), which would mount it on the top of the FSS. The RMS would then detach the EUVE payload from the EP and place it on a temporary mount on the FSS, as shown in Fig. 13. Then the RMS would detach the XTE PM from the FSS (Fig. 14) and mount it onto the EP (Fig. 15). The EP/XTE would then be released into space. The EUVE would then be securely mounted on the FSS in the Shuttle bay and returned to Earth.

I like to think that by publishing these sketches, I helped make clear that this was quite a complicated and potentially risky deal. It sure made that point clear to me. I wasn't so sure it was any more risky than other Shuttle related tasks, but I did want people to recognize it for what it was.

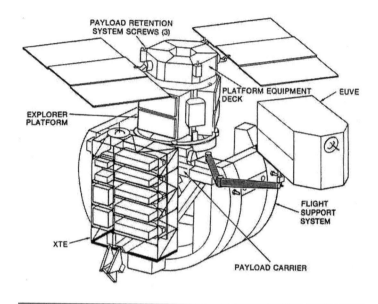

Figure 13: EUVE/XTE changeout in Shuttle bay, #1

Intermediate stage of the change-out of EUVE and XTE PM (Payload Module) in the Shuttle Bay during flight. The Flight Support System (FSS) fits crosswise into the aft end of the Shuttle Bay. The XTE PM (labeled XTE) is still in its launch configuration. The EP/EUVE had previously been captured from orbit and placed on top of the FSS by the Shuttle's Remote Manipulator System (RMS). The EUVE science payload had then been removed from the Explorer Platform (EP), and placed on a temporary mount to the right, as shown here. The EP, with its Platform Equipment Deck (PED), which holds the extended solar panels, remains on top of the FSS ready to receive the XTE payload. [Drawing by Goddard Space Flight Center. Presented at COSPAR, The Hague, July 1990 and published in Bradt, Swank, and Rothschild, Adv. Sp. Res. 8, 243 (1991)]

At one point, during the long interval of several years when this was the unquestioned way the mission would be carried out, the topic came up in a conversation with Riccardo Giacconi. He asked me why I had allowed such a seemingly far-fetched plan to go forward. After all, Riccardo was one who could tell the

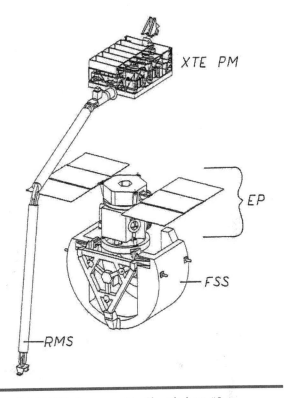

Figure 14: EUVE/XTE changeout in Shuttle bay, #2

Later stage of the change-out. The XTE payload module is being moved to the Explorer Platform (EP) by the Shuttle's Remote Manipulator System (RMS). EUVE is not shown. For references and definitions, see the caption to Fig. 13.

government how to do something, not the other way around. Perhaps I should be faulted for that, but on the other hand, this was Charlie Pellerin's way of keeping the mission alive when there was no money for it. XTE would ride the space-station craze while it was popular and get off it when Shuttles started blowing up and when the costs of being in the manned program became painfully evident.

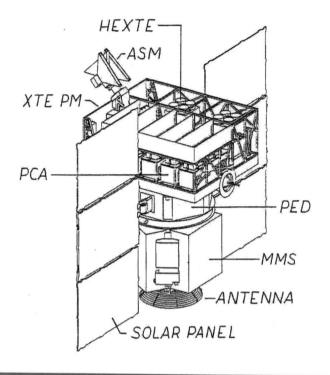

Figure 15: EUVE/XTE changeout in Shuttle bay, #3
The XTE payload module on the Explorer Platform (EP) in its final configuration as it would be inserted into orbit for observations. The core of the EP is a Multi-mission Modular Spacecraft (MMS). The Platform Equipment Deck (PED) supports the XTE payload and the rotatable solar panels. For references and definitions, see Fig. 13.

Later that year (1990), Jeff McClintock wrote a letter to Charlie Pellerin making the case for XTE in view of threatened cuts to the Explorer budget. The Explorer Platform preparation for EUVE was having difficulties also. EUVE would not launch until 1992, at which time a full decade would have elapsed since NASA selected the three XTE instruments.

Dedicated Spacecraft approved, July 1991

In the early months of 1991, NASA Headquarters was in the midst of deciding whether XTE would follow EUVE on the Explorer Platform or whether a dedicated spacecraft was preferable. Charlie Pellerin was now pushing the dedicated spacecraft option, despite the superficially higher cost, and of course, the XTE scientists were fully behind it. I wrote letters to the Goddard director, John Klineberg, and to Len Fisk, Associate Administrator for Science at NASA Headquarters extolling the value of that plan. In July, Goddard made a presentation to Fisk on the two options, and later that month he endorsed the dedicated spacecraft option. This was a huge win for the XTE concept. Our dream mission was coming to life!

It turned out that Ed Stone of NASA's Jet Propulsion Laboratory also played a role in the decision by writing a positive letter to Fisk. Ed was an investigator and the mission scientist on the *Advanced Composition Explorer* (ACE), which followed XTE in the queue. It is possible that ACE was scheduled for the EP also. His support of a dedicated spacecraft was very important for XTE as well as for ACE.

Ed was a cosmic-ray researcher whom I had known quite well when we both were working in the HEAO-1 program in the 1970s. Since then he had risen to quite eminent heights through his directorship of the Jet Propulsion Laboratory and his visibility on the Voyager satellite and Keck telescope programs. I ran into him at the Mauna Kea Hotel on the Big Island (Hawaii) during our time there for the total solar eclipse of July 11, 1991. I was the hotel eclipse lecturer, and he was there with a group of Keck telescope donors promoting the idea of a second Keck telescope. His letter to Fisk apparently followed this visit, but I had never once mentioned the issue to him at the Mauna Kea.

We had chatted once or twice but never had a one-on-one talk. His support, which was very important, was totally independent of any machinations by me.

The successful viewing of the solar eclipse and NASA's spacecraft decision made July 1991 a banner month for me. Morale among the XTE teams jumped to new heights. We had a spacecraft to die for (if you were an X-ray timing enthusiast) and three instruments under contract; we were ready to go. I sent off a thank-you note to Charlie with my heartfelt thanks, with copies to others whom I had been pestering. Again, he saved XTE from extinction. Of course, everyone involved deserved credit, the scientists, engineers, managers, and administrators at Goddard, MIT and UCSD. All helped to put together a viable mission within reasonable cost guidelines. But to my thinking, Charlie deserves first mention.

We were all set to charge ahead toward launch, but the broader astronomical community was less familiar with the role XTE would play in astrophysics. Accordingly, we scientists prepared a glossy 44-page document to distribute at the January 1992 AAS meeting to describe the mission and the science it could address. Was this another of Steve's ideas? It may have been. Several of us worked hard on it, and it was laid out and printed at MIT. Many hundreds of copies were distributed at the meeting. Its cover is reproduced in Fig. 1.

Roundabout data retrieval with TDRSS

Our goal for data retrieval was, as we have said, to have a nearly continuous telemetry link to RXTE through NASA regardless of spacecraft orientation (determined by PCA target and sun location) and position in its orbit. The plan was to use the two NASA relay satellites (TDRSS), which were in geo-synchronous orbit at altitude 5.6 Earth radii. This was a long way up, given that XTE would be orbiting at an altitude of only 0.1 Earth radius. Our data from XTE would be broadcast up to TDRSS from whence it would be relayed to White Sands, New Mexico, and then to NASA, and thence to the scientific investigators. Two of these relay satellites covered almost the entire track of the XTE orbit, thus providing near continuous coverage.

Commands to XTE would travel the same path in reverse. The large transmission distance to TDRSS required XTE to carry powerful transmitters and antennas with good gain, which added cost and complexity to the mission. It would seem to be a lot easier to simply broadcast directly down to the Earth's surface. However, since there would be only a few ground stations along the orbital path, the downlink time would be quite limited. Figure 16 is a not-to-scale sketch of the data and command path.

Instrumental challenges

In the meantime, the three instrument teams and the spacecraft engineers were at work on the design and then the fabrication of the spacecraft and the instruments. Typical of many engineering/scientific projects, difficulties were encountered from time to time. They could threaten the viability of an instrument and possibly slow the project and thus raise costs. In the end, solutions were found, though sometimes accompanied by a lot of anxiety. All three experiment groups and also the spacecraft team suffered such trials. Fortunately, none of the incidents proved to be showstoppers. The individual groups did not tend to share their difficulties widely but preferred to suffer through them quietly as they worked to solve them. I can only

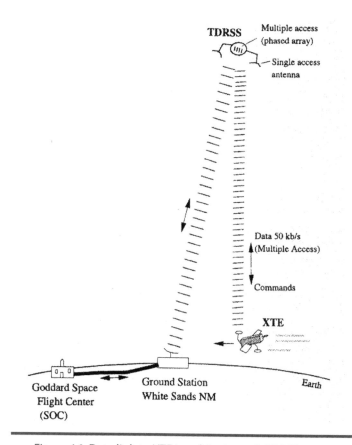

Figure 16: Data link to XTE in orbit, through TDRSS
*Path taken by data from XTE to GSFC. The scale is distorted: the TDRSS altitude is about 5.6 Earth radii, which, relative to the illustrated curvature of the Earth's surface, is about **eight** times the distance shown. The altitude of XTE, 1/13 Earth radius, is shown at about its actual height for the shown curvature. The transmissions between White Sands and Goddard may have made use of commercial communications satellites. This is how we got the data during the actual mission. [From XTE 1992 brochure. Sketch by H. Bradt]*

give testament, fairly, to some of the highlights, problems and lessons learned by our group at MIT. Here are a few of them:

The ASM counter deception

Finding a worthy vendor for the proportional counters that would serve as our ASM detectors was not a simple matter. It was not like buying a toaster with many qualified vendors out there. The counters were gas-filled aluminum boxes perhaps six inches in length with delicate beryllium windows (to admit the X-rays) and delicate quartz fiber anodes coated with carbon to which a high voltage (1500 volts) is applied. The gas absorbs the X-rays, which give rise to electrical pulses that can be recorded. A proportional counter is a more sophisticated version of the radiation-detecting Geiger tube.

We had planned to have these counters made by LND, Inc. on Long Island, New York. They had made counters for most of our prior missions and had always come through with reliable detectors, although often with unnerving schedule problems. It was a small firm with a strong commercial business building simpler counters. Our counters were much more challenging—they had to record the incident position of each detected X-ray—and hence were of interest to their engineers. But their highest priority would still be their commercial business.

We therefore decided to ask LND to build us a prototype detector; it was the only company qualified to do so as far as we knew at the time of our proposal. But as we commenced design work on the ASM, we became aware of a small branch of the large mining company Outokumpu in Helsinki, Finland, that made proportional counters with position-sensing and spectral capability for use in the mining industry, but again their counters were substantially simpler than our needs.

Because of concerns about LND's performance, our project manager Bill Mayer and project engineer Bob Goeke decided to explore the capability of the Helsinki company, which became

Metorex when it later split off from its parent. Its senior engineer, Seppo Nenonen, was interested and felt the firm was up to the job. Bob Goeke already had a family hiking vacation planned for Austria, so he took a side trip with his son to visit Metorex and came away convinced it had the required management and technical capabilities. A couple of months later our chief quality manager, Brian Klatt, made a special trip and came away similarly convinced that their quality control capability was up to NASA standards. It was thus decided to ask Metorex to provide a prototype counter in parallel with LND. We would then evaluate the two units and the performance of each company before choosing which firm would make the flight units.

When this plan was presented to the XTE project manager at Goddard, Dale Schulz, he objected on the reasonable grounds that we would be much better off focusing our attention on LND rather than spreading our oversight efforts across two firms, especially with one as distant as Finland. In that way, we could better monitor progress at LND and solve problems promptly. Also, building a second prototype was quite expensive, $60,000 including documentation and partial test units. We argued that the risk of depending on only one vendor was huge, and that our allocated funds would cover the cost. But Dale held his ground and directed us to stick with LND, and in most cases this would have been the last word.

At MIT, Bill Mayer felt strongly that, given our mixed experience with LND, we had to try both vendors. He knew that, according to our contract, it was our prerogative to decide such issues; this was a scientific investigation, not a procurement. He also felt that a verbal battle with Dale was not worth it, especially if we ended up choosing LND to build the flight detectors. So, he ordered the counter from LND and also one from Metorex without notifying Dale of the latter. We managed to keep the purchase under the radar on our monthly financial reports by paying for it through an American affiliate of Metorex. This uneasy deception carried on for six months or so during which Metorex made steady progress and LND missed deadline after deadline.

Finally, when the Metorex counter was delivered and we had selected Metorex to do the flight counters, it became necessary to tell Dale. We asked his liaison to MIT, engineer Jenny Glenn, whom we had also kept in the dark on this, how and when we should tell him. Her answer was, "By phone and quickly" because, by this time, she said, Dale was losing sleep and was practically physically sick about the lack of progress at LND, which could jeopardize our entire instrument. Needless to say, he was most happy to hear that we had a successful (almost) prototype. He did ask Bill, though, "What else don't I know that I should hear about?" Bill assured him that there was nothing else, and there wasn't.

This subterfuge meant that the travel expenses for those first exploratory visits to Metorex by Bob and Brian could not be charged to the NASA contract; Dale would never have approved the foreign travel. Bob thus bore the cost of his side trip to Helsinki himself. Similarly, Brian covered the cost of his entire trip. He recouped in part by turning it into a family vacation. As Bob recently wrote me, "When the group of people working on XTE thought something needed to be done, it got done." He added, "We were pushing the limits of going against 'direction' in our Finnish escapades, but it wasn't actually illegal!"

Bob recently commented on engineer Jenny Glenn's role as NASA's liaison to MIT: "She saved us untold grief in our relations to Dale and the Goddard management team and was instrumental in the smooth integration of the ASM and EDS onto the spacecraft at Goddard." Women were well represented on our senior engineering staff: Dorothy Gordon was the digital electronic engineer who designed the entire EDS; Ann Davis was the software engineer for the XTE ground support equipment.

Detector issues

The prototype counter from Metorex unfortunately exhibited a mechanical problem that was revealed in a "shake test" wherein a piece of hardware is tested for its ability to withstand the intense vibrations of a spacecraft launch. The connections from the high voltage feedthroughs to the inner frame failed. A design change was incorporated into the flight counters that were by then under construction. When we received the first ones with this fix, we did not immediately shake and thermal test them because such tests were scheduled to occur after we had mounted electronics on them, perhaps a year hence. We had confidence in the new design and doubly testing the counters seemed unduly risky. The ultra-thin beryllium entry windows of the counters and the filament anodes inside them were so fragile we were always nervous about them. Anyone who has seen a shake test would understand; the instrument components shake like a hummingbird's wings with a tremendous roar. It is frightening! You hold your breath and pray the test will stop before anything breaks.

Unfortunately, it was in the testing a year later that we discovered our design change had an unintended flaw that revealed itself in the thermal test. Finding this out at that late stage seemed to be catastrophic—a showstopper—since the flight counters had been delivered and were welded shut. Happily, Metorex saved the day. They had designed the counter with extra material at the weld, so they could mill off the weld, open the counter, fix it, and re-weld it shut, and repeat this several times if necessary. I love engineers who think ahead like this! In this manner, a successful fix was implemented. However, I wondered if our (or at least my) fear of an early additional test had been misguided. Perhaps, my fear of bad news, finding a flaw, inclined me to accept, more readily than I should have, the argument that an extra set of tests could be physically damaging.

Occasional visits to Metorex were required, for example to inspect counters before they were welded shut. I went to Finland twice, once with Al Levine and another time by myself. We were having troubles with high-voltage breakdown in the detectors and were carrying out tests and developing new procedures at Metorex. One of these was a very careful inspection of the counter anodes for dust particles before welding. For a couple of days, I sat for hours scanning along the anode wires with a microscope. Be assured that I, a rank amateur, was not the primary scanner, but rather was a backup gaining a feel for the process.

One counter had had a breakdown that had gone on for some hours before it was noted, at which point the high voltage had been shut off. I noted that the carbon resistive coating on one of the fine filamentary quartz anode wires in that counter had been burned completely off along perhaps a third of the anode length. Apparently, the arcing had worked its way gradually along the wire, burning off the carbon as it went, until the voltage was shut off. This observation stood us in good stead later when, after launch, we experienced several potentially catastrophic high-voltage breakdowns.

Another problem we had were pin-hole leaks in the counter windows. As we prepared for launch, one of the three counters on the flight instrument developed such a leak. Losing gas affected the counter gain. This could shorten the useful life of the detector, but this leak was so very slow that the counter would remain useful for several years, which was barely tolerable. At MIT, I pushed hard to complete testing on a replacement counter that had no such leak just in case the opportunity to access our instrument became available, though the chances were slim. These tests were being carried out when the entire XTE was at Kennedy Space Center waiting to be mounted on top of the Delta rocket or was actually already on it. As expected, we never got the chance to replace that counter.

The gas in the counters was just a bit above atmospheric pressure, so while still at ground level, the xenon gas inside the leaky detector was ever so slowly leaking out, but also some atmospheric gas (oxygen and nitrogen) was getting into it. This slow exchange took place continuously until launch, perhaps a full year. This unexpectedly had a good side. The oxygen in the counter slightly "poisoned" or *quenched* the gas, and this made it less likely to undergo breakdown. That counter turned out to be the most stable of the three after launch; it suffered no breakdowns. It was fortunate we never replaced it.

Of course, after launch, that leaky counter was still leaking, but now that it was in the vacuum of space, it was no longer being poisoned by atmospheric oxygen. Nevertheless, the xenon gas very, very slowly continued to escape. The slow descent of the gas pressure in the detector caused the output pulse heights to ever so slowly increase over the following months and years. Recalibrations in our data processing easily compensated for this for many years. In the latter part of the mission, though, the counter required increasingly difficult recalibrations, until toward the very end of the extended mission, it became essentially useless. But it did provide useful data for almost all of the 16 years of flight operations!

Radioactive principal investigator

I had open heart surgery (quadruple bypass) in the late summer of 1995—the year of the December launch—and recall that one of the trips to Finland was shortly after that. My attitude toward taking this trip so soon after the operation was quite in contrast to my attitude shortly after my first heart operation, an angioplasty in 1988. At that time, I felt terribly vulnerable and wanted to set up camp just outside the emergency room at Mass General Hospital. This time, the fast-approaching launch kept my mind focused on XTE, not on myself.

During the medical testing prior to the bypass operation, I had had a stress test with a radioactive tracer to diagnose my coronary circulation. When I was loaded up with radioactive material, I went to our lab at MIT where we had an ASM detector on long-term test to see if breakdown would develop. This was the detector I wanted to be ready as a flight replacement should the opportunity arise.

When I arrived, I grabbed a student who happened to be working there and also Bill Mayer, telling him, "This should be interesting." I took them into the room where the detector was under test, slowly clicking away from the natural radioactivity of the room and cosmic ray hits. The detector count rate remained low as I stood just inside the door. Then I took a few quick steps closer to the detector until I was perhaps six feet from it, at which time the counting rate indicator jumped way up, as I knew it would. I immediately stepped back, and the rate dropped to normal again. Bill said, "Wow, that was scary!"

I don't know if he was concerned about me because I was so loaded with radioactivity or because the count rate increase looked so much like a breakdown event that it could mean our replacement counter was defective. I think it was the latter. Figure 17 is the record of my radioactive visit on August 17, 1995.

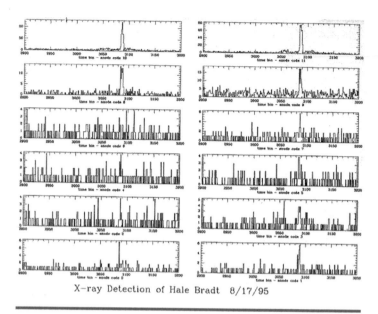

X-ray Detection of Hale Bradt 8/17/95

Figure 17: Detection of radioactive ASM principal investigator
Detection of radioactive Hale Bradt with one of the ASM test detectors. The twelve panels represent the counts from the several different anodes, or sets of anodes, of the detector. Time runs from left to right for 300 s. All rates jumped up for the several seconds when I was close to the detector.

ASM software

Ron Remillard recalled for me the ASM software crisis. The programmer working on the analysis software for the ASM data had designed the tasks with complicated C++ structures, giving few milestones by which progress could be measured. Then some three months before launch, he quit, with very little or nothing to show for his time on the job. This was another potential catastrophe. Ron was able to bring to bear his own prototype analysis software, which he had written to process test data for his own purposes. Features were added later by scientist Wei Cui who had joined our group and even later it all was upgraded and improved by Alan Levine.

Ron credited the availability of his software to a conversation with Gunther Hasinger, a lead scientist on the German *Rosat* mission. He had told Ron about a similar crisis regarding their analysis software and how they had used Gunther's prototype software. Gunther's strong advice to Ron was that, during the implementation phase, keep your prototype software up to date, so it can serve as a backup, and Ron had done so.

Why doesn't this crisis loom large in my mind today? I frankly had forgotten about it. Al and Ron had taken over the responsibility for the ASM so effectively that I had relinquished much of the worry to them. For example, they realized the shortcomings of the analysis scheme for ASM data in our original proposal and initiated the more straightforward multiple dwell operation of the ASM. They also made an extensive study of detector mask patterns and introduced different patterns for the several segments of a single mask, thereby providing positional information normal to the high-resolution direction. (It could reduce the length of a source's line of position.)

Failure of the EDS

The EDS (Experiment Data System), recall, was the electronics box that processed data from the PCA and ASM instruments before sending it to the ground via telemetry. As described above, it was essentially a computer consisting of ten circuit boards of which eight were Event Analyzers. Each Event Analyzer (EA) was a little computer with its own Digital Signal Processor (DSP) chip and associated circuitry. We had never built a flight computer, and much thought was given as to the best way to end up with a robust reliable system.

We began with a requirements document that listed all features desired by the scientists, each with levels of desired performance (minimal, nominal, best). This gave the engineers the understanding of our needs and the flexibility to meet them. This document was negotiated with the engineers twice before design began. Ed Morgan, the EDS instrument scientist, kept in constant contact with the engineers at both MIT and Goddard throughout its development, so misunderstandings or problems were quickly resolved. The result was that there was very little if any escalation of requirements during the design and fabrication phases, and the final product performed as expected, except for one huge problem, which I now recount.

The completed EDS went through all its functional and environmental testing successfully under the watchful eye of our engineers and Ed Morgan. Its delivery and integration into the spacecraft at Goddard in Greenbelt, Maryland, was scheduled early in the integration, as it had to be in place to accept the signals from the ASM and PCA when they too were installed into the spacecraft. The spacecraft with all instruments installed would then undergo its own functional and environmental tests.

The assembled EDS was a beautiful rectangular case full of electronic circuitry (Fig. 7) with connector ports on the ends. It was completely anodized so it had a golden color and also a big MIT logo on its side. We were quite proud of it. That logo was of some consternation when the EDS was received at Goddard because it was felt that it should have had NASA's logo on it, not MIT's. After all it was a NASA procured item, no? Well, not really; this was MIT's *scientific instrument*, not a NASA widget.

The EDS was to be delivered to NASA/GSFC, on a Monday. It was in the lab, connected to a computer so that staff, in the lab or remotely, could communicate with it. Late Friday, Bob Goeke, started it up to do a last-minute operational check before unplugging and packing it for shipment. To his horror, he found one of the event-analyzer boards would not boot (turn on). This was big trouble that would have to be addressed after the weekend. He left it hooked up so further diagnostics could be carried out Monday.

Unaware of this on Monday morning, Ed Morgan decided to give it a quick check before coming into work. It was his baby, after all, and he was wanted to do his own last check. He was in the habit of running software tests on it whenever he could and had found a number of software bugs in the course of thousands of tests. He logged in from his home and also found that, to *his* horror, one of the eight Event Analyzers (computers) in the EDS would not boot. This was a disaster; it had been extensively tested and was to be delivered to NASA that very day!

This failure produced lots of angst and led to intense diagnostics. It was quickly determined that an electrical connection on one of the internal layers of a multi-layer circuit board had failed. The defective board was shipped to MIT's Lincoln Laboratory for further analysis. Was it a minor isolated failure or was it endemic to that entire circuit board or to *all* the boards in the EDS? We did not know. The EDS was essential for processing all of the ASM and PCA data during all the forthcoming environmental tests of the entire spacecraft. This was a big scary deal that attracted lots of attention.

Large programs like XTE spend money like water in the later stages of the program because so many people at so many institutions (universities, NASA, and subcontractors) are actively working at full bore. That spending continues during any stretch-out of the schedule due to a holdup in any one unit. Hence there is tremendous pressure to stay on schedule while resolving problems. In our case, the EDS was key to the entire mission; the principal instrument, the PCA, was useless without it. Thus, it threatened a costly delay; it could not be left off the spacecraft as could a secondary science instrument. "Sorry, you missed the bus."

That same day or possibly the next, we were huddled in Bill's office reviewing our options. The Goddard project office was desperately trying to reach us, and we did not want to be reached. As I recall, Dale Schulz at Goddard was concerned that we might take steps that would jeopardize the integrity of the EDS, which he needed for the spacecraft integration. He was objecting to us sending the board to MIT's Lincoln Laboratory for diagnostics, before more thought and discussion with Goddard, or perhaps because he wanted Goddard labs to do the diagnostics. We had already sent the board to Lincoln and were focused on getting to the heart of the problem as fast as possible. Dale's directions might well interfere with our efforts. We had the EDS, which had not yet been delivered. We believed we owned this problem.

So, when the phone rang in Bill's office, he did not answer it, saying he knew it was Dale. Then, we could hear the phone ring next door in Bob Goeke's office, and it wasn't answered. The phone could be heard in several other offices too, and they also went unanswered. A bit later in our meeting, one of our engineers, Ann Davis, came down the hall, stuck her head in the door, and said, "I just got a call from the project manager at Goddard who says he can't reach any of you. What's going on? I never get called by him." Bill said, with a twinkle in his eye, "Why did you answer it?" Her reply: "You didn't tell me not to." Such was life in the Albert Hill building in Kendall Square, Cambridge, Massachusetts, where all this was taking place.

It turned out that the problem was indeed endemic to all the circuit boards in the EDS. This was the worst possible news. These boards were *multilayer boards* with as many as five or six layers of board material sandwiched together, each with its little metal traces running here and there between the chips mounted on the top of the sandwich. The optimum way to solder them is to mount all components on them, heat the entire board and components to an appropriate temperature, and then to pass the board through a "wave" of molten solder. We had not used this method because of the delayed delivery of some components.

To maintain schedule, we had hand-soldered the components as they arrived, the earliest ones first. One location on the boards had a large ground plane on one of the inner layers, so additional local heat had been applied locally to the nearby solder joints. This led to unseen damage to the board material that in turn led to internal broken leads. Or so we thought at the time.

A contributing factor was that the board material was different than that with which we had had prior experience. The new material (polyimide) was recommended as superior by Goddard, but it was not superior in this one respect, namely its susceptibility to damage by extra local heat while showing no visible symptoms. We and perhaps Goddard were not aware of that shortcoming when we decided to hand-solder the components. The issue probably was not on anyone's radar because standard practice was to use wave soldering. Is there a lesson here? Use standard practice whenever possible? Perhaps, but that could quench valuable innovation. On the other hand, for space hardware, too much innovation can be dangerous.

That is the story as I remember it, but Bill Mayer recalls that the problem was eventually tracked to a board fabrication issue. The vendor had not adequately cleaned the individual layers before lamination, and residual etchant chemical (acid) etched away conductors (traces) with time. So perhaps our soldering procedure was less at fault.

This type of issue arose more than once. When buying a product for a space mission, one looks for the qualities one needs, such as performance in a vacuum or under temperature extremes. Goddard's quality department is one resource of such information because of the wide variety of space missions in its history. However, the catch is that they might come up with a list of approved materials from a previous mission where the needs were somewhat different. Hence a "better" material or item might not be better for your particular use.

For example, we needed the most reliable lubricant for the rotation bearing on the ASM. It had to be effective for years in the vacuum of space. We sought out opinions from industry experts, as well as from NASA, and found that some of the more credible recommendations were totally opposed to one another. The resolution of this issue finally came down to our assessment of the sources of the recommendations. In the end we followed that of a vacuum lubricant expert at Ball Brothers, a company well experienced in space missions.

The Great EDS Switcheroo

The big question before us now was what to do about the failure of the EDS circuit boards; there was no spare EDS. To rebuild the boards would take months, and this would delay the integration and testing of XTE. Such a delay would invoke huge additional costs, not to mention a major loss of face. An ingenious solution—though obvious in hindsight—was found; I believe it was Bill's or perhaps he and Bob concocted it together. They always ate lunch together in Bill's office, and most of the difficult issues facing them were cogitated on, or should I say chewed upon, between bites of sandwiches.

The plan they came up with was to install a jumper on the

defective circuit boards at the locations where the broken lead problem had arisen. The existing EDS with the patched boards would then be delivered for integration and for testing of the spacecraft, which could then go forward on schedule. After the testing, the defective EDS would be removed and replaced by an entirely new unit that MIT would build in the meantime.

This was the "great switcheroo." It involved obtaining and then assembling not only new boards, but also all the board frames, connectors, and wiring. The completed totally new EDS would then undergo testing somewhat more rigorous than the usual practice to compensate for the tests it would miss aboard the spacecraft. All this was more work and more cost, but it would keep the program on schedule. We were proudly able to accomplish this without, in the end, increasing our Total Estimated Cost, as described above. Again, budgeting for unknown unknowns paid off.

Over the next several months, our team ordered or fabricated parts and assembled them into a completely new EDS. Replacement circuit boards were already in hand. We had initially ordered a set of circuit boards from two different vendors; the unused set was qualified and still in storage in our laboratory. The electronic components and chips were installed on the boards, which were then wave soldered. The completed EDS then satisfactorily passed all environmental tests at MIT, which were somewhat extended to account for the spacecraft tests it had missed. It then was installed on the spacecraft and performed flawlessly in orbit for the entire 16-year mission.

The defective unit had gone through the spacecraft environmental test successfully despite having patched-up boards. It had done its duty and was returned to our laboratory (Fig. 7), where Ed used it to test EDS software patches for many years during the RXTE flight operations.

Procurement near misses

We learned a few things in the process of building the second EDS unit.

For example, there was a problem with getting enough memory chips for the boards. For the initial build, Goddard had made a bulk buy of one particular type that was *radiation hardened*—for protection from cosmic-ray hits in space—so the several instrument teams could benefit from the lower cost of the large purchase.

Our project manager, Bill Mayer, asked Goddard how many they had left that could be used for our new EDS. They had a significant but insufficient number. Bill still needed several dozens, say 100, more of them, so he called the manufacturer asking if they had the item, referring to the part number printed on the case of one of the units from the first batch. They replied that yes, they did. Bill asked if they had 100, and they replied that yes, they had several thousand. Bill then asked what each would cost, and they quoted some very low number as space qualified parts go, like $20 each. (In the spacecraft hardware world, that is like two cents.) That certainly seemed like an easy cost-effective solution.

Bill then informed the Goddard Quality office, as required, of our plan to use these, and they informed him that we could not use them; they were not radiation hardened, which was a necessity for space use. Bill protested that the $20 items had the same serial number on the case as those that Goddard had provided us for the first EDS. He was informed that in preparing the bulk order, Goddard had obtained units with that part number but had then had them hardened elsewhere without changing the number on the case! What a recipe for confusion! Needless to say, we did not buy the $20 units. Bill managed to locate just enough odd depositories of the qualified memory chips to fill his needs.

Another parts problem revealed itself during the fabrication of the new EDS; it involved the Digital Signal Processor (DSP) chips. This chip was the engine or heart of our Event Analyzer computers. We notified the Goddard Quality office that we would use the same item, specifying the part number, as on the first EDS. The response was that this chip was not protected against "latchup," a catastrophic short circuit that destroys the chip due to a cosmic ray depositing ionization energy in it. We thus had to design protection circuitry for the DSP on each of the eight new boards. Without this fix, we estimated that we would have lost roughly one of the eight boards a year. Shockingly, the original EDS had this problem, which was unknown to us until we were building its replacement!

We looked back to see how this use of an unqualified part had occurred during the initial fabrication. It turned out that we had sent a list of EDS parts to Goddard for approval. This was required and also necessary because Goddard had information about quality issues not available to us. Approval of a document by Goddard was often delayed because the document had to pass through a number of cognizant offices, and this could lead to large delays. Often though, our engineers would learn essential information informally. In this case, we had received an informal response indicating that two of the items on the parts list were not qualified for space flight. We took this to mean that the other items were qualified, and the (unqualified) DSP chip was among them. We then proceeded to use them. We had never received notice that they were not qualified.

This was a continuing problem we had with Goddard (from our perspective). Word-of-mouth agreements created a risk of failures or that interfaces would not match at integration. Bill felt this was jeopardizing the program, not only at MIT, but elsewhere in the XTE program. In a lengthy, strong, and detailed letter, he urged Dale Schulz's boss, Jim Barrowman, to solve this documentation problem. I am not sure if there was ever a clean resolution of this, but fortunately, there were no other major miscommunications, to our knowledge. There apparently were enough informal discussions between the MIT and Goddard engineers. But this story illustrates the value of efficiently managing the formal documentation.

An interesting quid pro quo developed between Bill and Dale. Although most of Bill's interactions with Goddard were in person or by phone, he would write a letter if he felt something needed serious attention. His letters, typed by the group secretary, were clear, long, full of facts, and usually right on the mark. Sometimes, they would be strongly stated and justifiably so. Dale was the usual recipient and sometimes referred to them, with a chuckle, as "Bill's nasty-grams."

Dale made a point of never answering these in writing, but rather would phone Bill, so misunderstandings would not escalate. Dale was good at staying calm when things were tense and that helped. Bill could flare up (as could I) but would quickly return to business in a constructive friendly manner. His great competence and knowledge of the program were huge pluses. He could see and communicate the big picture, and this helped discussions get over rough spots.

On one occasion, during one of our obligatory trips to Goddard, Dale invited us and other XTE attendees to his home on the Maryland shore for a drink, and then we all went out to supper together (Dutch treat). He did it deliberately to enhance the interpersonal relationships. I thought that was especially nice and always meant to return the favor when Dale and his team were in Cambridge, but I never quite got around to it. I still regret that.

Management issues

Before launch, scientific input was provided to NASA by a group of scientists known as the XTE Science Working Group. It included the instrument principal investigators as well as several outside scientists. It may have begun meeting in 1991. Membership evolved with time. Dan Schwartz was the first chairman. Upon launch in December 1995, the committee became the RXTE Users' Group. In 1997, Fred Lamb took over as chairman and continued until May 2008. John Tomsick then served in that role until the end of the mission in 2012. Other scientists serving at various times in these groups included John Deeter, John Grunsfeld, Paul Hertz, and Shri Kulkarni.

The committee would meet, perhaps twice a year, and would serve as a sounding board on a wide variety of issues. It was particularly attentive to issues affecting the wider user community. Michiel van der Klis of the University of Amsterdam in The Netherlands was notable for his participation in these groups and his continuous enthusiasm and support throughout the RXTE program.

During the mission, spacecraft operations would be managed by several groups at Goddard, led by scientist Frank Marshall, director of the RXTE Science Operations Center (SOC). He and his colleagues, notably scientists Alan Smale, Robin Corbet, and Padi Boyd, would create observing schedules and monitor spacecraft and instrument performance. They also would provide support to the observers, processing proposals and providing data analysis tools. The entire effort was part of Goddard's HEASARC (High Energy Astronomy Science Archive Research Center), which archives all satellite data from NASA's high-energy astronomy missions. Preparing for all this was a major effort with its own challenges. Like the hardware, all this had to be up and functioning at launch.

In 1994, as in-orbit operations were being planned, I raised an issue that caused some consternation at NASA Headquarters and also among some of my colleagues on XTE. Since the mission was to be 100% guest observations, it became apparent to me that the principal investigators of the instruments could well be relegated to bystander status in the scientific operations and finances of the mission. The Users' Group would include the principal investigators (Jean Swank, Rick Rothschild, and me) as members, but we would be a minority, and the chairman would not be one of us. NASA policy as stated in writing by Guenther Riegler of NASA Headquarters was, "The NASA Project Manager and Project Scientist have ultimate responsibility for all decision-making and implementation, including directing all mission operations and data analysis."

As a counterbalance to this, I proposed a Mission Management Working Group consisting only of the three principal investigators, with ex officio members being the RXTE Science Operations Center director, the NASA project scientist, and the chair of the RXTE Users' Group. Its purpose would be to:

1. Advise NASA (Headquarters, Project Scientist, Project Manager) regarding
   Operations priorities
   Implementation of policy, short and long term
   Implementation of recommendations of User Committee
   Allocation of resources.
2. Propose new policy as needed.

Meetings would take place only as needed, frequent consultations could take place by telephone, and meetings would be open to non-members.

Looking at this now, it seems eminently reasonable; it would be a kind of executive board of the RXTE Users' Group, empowered only to give advice. But NASA Headquarters and also Jean Swank, NASA's XTE project scientist, viewed it as unnecessarily cumbersome. I was assured that NASA would listen to the principal investigators without such a group. Since my proposal was received with such dismay by Headquarters and since my colleagues had so little enthusiasm for it, I let it drop.

As the mission developed, Jean Swank, as project scientist and also principal investigator of the PCA, worked very closely with the other instrument teams. Goddard RXTE scientist Tod Strohmayer, who succeeded her in 2009, continued in that mode until the end of the mission. The instrument teams, knowing the mission capabilities so well, had a large influence on the User Group's deliberations. In fact, the RXTE Users' Group worked well as a single group with few or no serious disagreements. Thus, in the end, my concerns seemed unwarranted.

But in principle it could have turned out differently given different personalities and working styles. Jean and Tod, in fact, ran the mission as if my proposed working group existed, but she needn't have done so. I also think the objectors did not fully factor in the implications of going to 100% guest operations. Does that mean the principal investigators must give up having a separate voice in giving policy advice? If the conception of the mission was theirs, I think they should formally have an influential role.

Black holes at MIT

While all this was transpiring, I heard indirectly that a college roommate of mine, James Melchert, now a highly respected artist, would be installing one of his art works at MIT. It is a ceramic-tile mural named *Coming to Light*, 14 feet tall and 225 feet long (!) along one wall of a highly visible street-level corridor of the Koch Biology Building. It consists of a multiplicity of abstract images, some reminiscent of biological imagery and some geometrical in form. Several of the latter are simply solid black circles of varying sizes from about three feet in diameter to one larger than the 14-foot height of the mural and thereby partially truncated.

I believe those black circles represent astronomical black holes, in that they were subliminally in the artist's mind. (Jim does not admit to conscious awareness of this.) So, why do I think this? Simply because, when I met with Jim for the first time during the 1994 installation at MIT, his first question to me was, "Are you still working on the same thing you told me about on our last visit [five years earlier in Oakland, California, his home base]?" I responded, "What was that?" and he replied, "Black holes." Since XTE—all $200 million dollars of it—was charging ahead full-bore toward launch at that time, I could truly reply that, "Well, yes, I guess so."

Black holes and MIT, known to him primarily through me, had been in his head all along as he conceived and proposed his work for MIT! Case closed.

VI
Launch Attempts (December 1995)

Anticipation

THE READY-TO-FLY XTE SPACECRAFT with all instruments aboard arrived at Kennedy Space Center (KSC) at Cape Canaveral aboard a C-5 military transport jet on May 31, 1995. At that time, the launch was anticipated to be on August 31, but the date was under review. That date was passed by, possibly due to the launch vehicle (Delta II) not being ready or to scheduling issues at KSC. Our mission wasn't the only show at KSC.

In early October, XTE was ready to be moved to the launch pad and mounted on the Delta II. Before the move, on October 6, a science presentation and photo-op for the media took place at KSC. Rick Rothschild, Fred Lamb, and Richard Rothschild gave the science presentations. At that time, the launch was scheduled for November 6. That too was postponed.

The launch of XTE was finally firmly scheduled for December 10, 1995, and I remember being at the Cape on my 65th birthday, Dec. 7. I was there well before that because I also recall seeing from my motel balcony the spectacular launch of another NASA science satellite, SOHO, by an Atlas rocket, which took place on December 2. I have no idea why I would be there so early because the scientists are totally unnecessary at this point unless it is for last-minute checks of their instruments. It may have been that we were anticipating an earlier launch.

Just being at Kennedy Space Center gave me a visceral thrill, a sense of walking on holy ground, knowing the history of what had transpired there and was continuing in the present. The Shuttle program was then in full swing, with launches that November and January. I did watch one launch attempt that was scrubbed. Unfortunately, I was never able to see a Shuttle lift off.

By the time I arrived, the XTE payload had long been mounted on the rocket, a Delta II with nine solid-fuel boosters surrounding its base and an Aerojet second stage. I took part in the last visual inspection of the payload, high in the tower structure surrounding the bird, before the closing of the shroud that encloses the entire payload during the ascent. Would I be astute enough to note something out of order when I had neither prior experience nor a list of things to check, or was it simply a chance for me to see our baby one last time before it was projected 500 km beyond our reach?

It was mostly the latter I think, though I did remember the story of one of Columbia University's experiments being launched from White Sands Missile Range. (I wasn't there but

Figure 18: XTE and shroud in launch tower

XTE on top of the Delta II rocket with two thirds of the shroud that will surround it already in place. The five rectangular sunshades for the five PCA units and the two circular star-tracker sunshades are visible, the latter to the left of the lower PCA units. The ASM is on top of the box-like spacecraft structure. The three white panels are the back sides of the three SSC cameras. The HEXTE (not visible) is below the PCA. The large dark panel on the left is one of the solar panels, folded for launch. [NASA]

felt close to this as we also launched rockets from White Sands and once were there with the Columbia team.) The White Sands launch crew—a US Navy group—failed to hook up a wire from the tower to a valve on the rocket engine. When the solid-fuel booster started the rocket moving, the wire would, if properly connected, pull free allowing the valve to open, and this would start the liquid fuel rocket engine. In this case, with no connection, the solid-fuel booster got the rocket moving at high speed, but the liquid-fuel main engine never ignited.

After only two seconds of flight, barely out of the tower, the booster burnt out, its roaring sound stopped, and there was perfect silence. The load roar of the sustainer was expected to continue for another full minute. The rocket had no propulsion; it was in free flight. It rose for a mile or more full of unburnt liquid fuel like a German V2 rocket heading for London. It then returned to Earth nose first. The Columbia polarimeter in the nose of the rocket was not much to look at after the impact. If I had been there and made a last-minute pre-launch check of the tower, would I have noted the unconnected wire? Probably not, as I didn't even know about that wire until I heard this story.

On December 8, at the Cape, there was a Prelaunch Press Conference, for which I was the lone scientist. I was joined by XTE project manager Dale Schulz, the launch manager, the vehicle manager, and the weather officer. This was much more a rocket-launch presentation than one about science. Everything was still on track for a December 10 launch.

On the ninth, friends, associates, and families and also NASA officials gathered at KSC for the pre-launch festivities. I expected to have about 20 visitors, among whom were the large family of a cousin who lived in nearby Titusville, Florida. One of my daughters and a sister were there with their families as were another cousin, and an aunt and uncle who wintered in Florida. Since neither NASA nor MIT paid travel expenses for visitors, living nearby made attendance more likely.

NASA put on a wonderful show for the visitors the afternoon before the scheduled morning launch. A session in a small theatre featured talks by NASA officials and the principal investigators (Jean Swank, Rick Rothschild, and me), in which we explained the scientific goals of the mission. This was followed by our own private bus tour of the space center. That meant more or less the standard tour including the most impressive Shuttle facilities.

The highlight, of course, was a special stop at the Delta launch site where our rocket and payload were situated. Dusk was approaching and the white rocket, illuminated by searchlights and ready for launch the next morning, was a beautiful sight. Previously, staging had surrounded the rocket, so it was hardly visible, whereas now it was fully exposed under bright lights for all to see. Since the bus was full of XTE supporters and friends, it was a wonderful moment.

Earlier that day, photos were taken of various groups of XTE team members with the rocket in the background (Fig. 19). The whole day was a most festive time for me.

Attempts 1–4: high winds

The day of the launch, December 10, was clear but windy. The morning countdown proceeded as the various rocket and safety systems were checked. High-altitude wind velocities were measured with weather balloons launched for that purpose. High winds, either local or high altitude, led to several holds, until finally, the launch window was exceeded, and the flight attempt was scrubbed.

During the several hours of the countdown and holds, the visitors waited patiently in the chilly morning air on the bleachers set up some distance, a mile or two, from the launch tower. John Grunsfeld, an astronaut and MIT alumnus who had worked in our group at MIT, being a friend of XTE, came around in his space suit. I encouraged my five-year-old grandson Ben to

shake his hand and have a word with him, but the suit was too frightening, and he shied away from it. Also, my nephew Dale's wife fell and broke her coccyx that morning. All in all, it was not a good day.

Two more attempts were made the next two mornings, and both were aborted due to winds. Needless to say, the attendance in the viewing stands tapered off for those attempts, December 11 and 12. An Atlas launch of a commercial satellite was scheduled for the 15th, so the fourth attempt was delayed until the 17th, which was also scrubbed for weather.

These launch attempts were hard on the technical staff. They were at their stations when the countdown began the evening before a scheduled morning launch. After a launch was scrubbed, a series of steps had to be taken to ensure the safety of the rocket and to get all systems ready for the next launch attempt. Thus, some technicians would not be freed from their duties until noon or later. Then that evening, the whole process could start again. It was like the movie "Groundhog Day" where the protagonist is condemned to live February second over and over until he finally gets it right.

There were teams for every system: Delta rocket, second stage Aerojet motor, weather, range safety, science payload, and more. We never saw most of them as they were tucked into different locations at KSC or further afield. You knew they were out there as the launch director polled each of them on

Figure 19: Rothschild, Swank, Bradt, and Delta II
Principal investigators of the three instruments carried aboard RXTE and the Delta II rocket. From left, Richard Rothschild (UCSD, HEXTE), Jean Swank (GSFC, PCA), and Hale Bradt (MIT, ASM) on the day before the first launch attempt, 9 December 1995. The entire experiment and spacecraft are contained in the white larger-diameter shroud at the very top of the rocket as shown in Fig. 18. Nine solid-fuel boosters surround the 8-foot diameter base of the rocket. [NASA photograph].

their readiness at several stages during the countdown. We, the scientists, were not allowed into the main control room; only Dale Schulz with an assistant or two had a chair with computer screen in "mission control."

The instrument principal investigators were considered absolutely useless to this enterprise. Our instruments were not powered on, so there was nothing for us to evaluate for launch readiness. But we were allowed into a glassed-in VIP room that overlooked mission control. We could tune into various voice circuits to hear what was going on—the polling of the various teams and reports from the weather team, for example—but we could not talk into those circuits. I found it quite exciting to listen to the reports during the launch countdowns and also in the first minutes after launch; so many systems had to operate flawlessly.

I watched the launch of the astronauts in the SpaceX Dragon capsule recently, and the polling by the flight director was done by computer. The drama of hearing the spokesmen of the various teams' report, "Range safety, ready for launch," "Delta vehicle, ready for launch," . . . one after the other, is no more.

Attempt 5: main engine shutdown

I had returned to MIT after the third attempt, probably for academic duties, but also, we were trying to qualify a non-leaking detector as recounted above. The most nerve-wracking launch attempt occurred on the fifth attempt on December 18. The rocket had been on the launch pad for weeks, and condensation had collected on the valve that controlled fuel injection into the main engine. The loading of liquid fuels at subzero temperatures during the countdown froze the condensation and disabled the valve. Unbeknownst to all, it could not open. The countdown proceeded right down toward zero; a few seconds before zero, the small vernier (side) engines ignited as expected. But then, when zero was reached, there was no ignition of the main engine.

Ignition would have been detected, I was told, by a fine wire in the engine chamber carrying an electrical current until the engine flame vaporizes the wire. The loss of electrical continuity signals the existence of a flame. This in turn triggers the ignition of the nine solid-fuel boosters. Once the boosters are ignited, there would be no stopping the lift-off. In our case they did not ignite. If they had, it would have been the Columbia rocket fiasco on a much grander scale. It would not have been pretty. Igniting the boosters is like firing a gun—there is no stopping the bullet after ignition.

What a letdown!! Mission aborted at $t = 0$! Someone in the Mission Control yelled: " Oh s - - - !" That person was careful not to activate their own microphone, but it was picked up faintly by the microphone of the official NASA announcer, whose desk was in the back corner of the room. He had just depressed his mike button to announce the shutdown, so the expletive went out to the whole wide world. In fact, the outburst was only faintly audible, but the inflection made its meaning unmistakable. This led to some strong cautionary advice from NASA Headquarters later that day. Such outbursts did not match NASA's preferred image of itself.

I was not present at this launch attempt, as we were just getting ready to test the replacement ASM detector that morning at a test facility in Boxborough, Massachusetts. Of course, if the launch went well, we would not bother with the test. We took a brief break to call Ron Remillard who was listening to the countdown from his home in Boston. When he told us of the main-engine shutdown, we shrugged our shoulders and went back to our testing. Perhaps, if these failed attempts continued, the second stage rocket would require refurbishing and that might lead to an opportunity to replace the counter. The likelihood of that possibility was considered to be so small as to be laughable, but my colleagues humored me by continuing the testing. You never know . . . .

Rocket issues

The engine failure on the 18th required some post-mortem studies by the Delta team, which led to more delay and another review. In the meantime, several other issues with the rocket had come to light and were being studied. When it seemed that they had been resolved, the NASA launch director called a meeting of all concerned to review their status. It was held in a large room with 50 or more people from all the different involved teams. A high NASA official and the senior Delta team person from McDonnell-Douglas Aerospace hosted the meeting. They went through the several issues one by one. I had returned to the Cape and so was able to attend; principal investigators were invited.

This discussion was taking place in the context of tremendous pressure to get this mission launched. The previous launch attempts and delays were stretching budgets and people to the breaking points, and more delay could require refurbishment of the second stage Aerojet engine, a relatively big deal. On top of this, the US Government had been shut down since December 16 in a budget crisis, though the launch teams were kept on the job. These problems were made abundantly clear to all at the meeting. Getting a successful launch soon was a very high priority. Shades of Challenger!

The issues on the table were these:

Problem #1 – a spark

A senior respected technician was performing a task inside the main engine of the Delta rocket with pliers in his hand when he saw a flash or spark, perhaps jumping to the pliers. This was alarming but could not be reproduced. Perhaps he imagined it, but he was a highly reliable and experienced person. Since many other inspections by others and functional tests revealed no fault, this was not deemed sufficient reason to delay the mission.

Problem #2 – a washer

A small washer had inadvertently been dropped into a part of the main engine and could not be found despite an aggressive search. It was possibly lodged invisibly behind a gasket. In that position, it could conceivably lead to serious damage, but this was highly unlikely. This too was deemed to be an acceptable risk.

Problem #3 – potentiometers

Prolonged exposure to moisture on the launch pad caused a film to accumulate on the potentiometers that control the pointing directions of the two vernier engines. In a previous test, the film had rendered the controls ineffective. The verniers are the small thruster engines on either side of the main engine. Like the main engine, the nozzles can pivot in either sideways direction to effect pitch or yaw rotations of the rocket. Since they were located on opposite sides of the main engine, they also controlled the roll angle of the rocket. Diagnostics had determined that the film consisted of mineral deposits from the moisture in the atmosphere during the rocket's long exposure to the seaside air, or some similar natural effect.

Since the film had been removed recently, or the potentiometers replaced, it was deemed unlikely that new deposits could have accumulated sufficiently to cause problems on the forthcoming launch. Furthermore, the vernier engine motions would be tested in the last few minutes before launch. Also, vibrations of the rocket firing during launch were likely to shake off any accumulated deposits. It had also been determined that main-engine pivoting could apply sufficient torque to counter expected wind forces in the jet stream in case the vernier engines could not be pivoted. For these several reasons, this problem was considered well enough understood to allow a launch.

During the question period, a member of the audience asked, "If the vernier engines fail their pivoting test just before launch, would you proceed with the launch?" The NASA official at the podium in the right front corner of the room answered, "Yes, because it has been demonstrated that the main engine has sufficient torque to traverse the jet stream successfully." The clear sense was that NASA just had to get this rocket off the ground so it would stop bleeding money. While he was saying this, the spokesperson for the Delta rocket team in the left front corner at his podium was busily shaking his head from side to side unnoticed by the NASA official. When he had a chance to reply, he said simply, "We do not launch with a known failure." That vividly illuminated the different perspectives!

Finally, the conclusions of the several studies were summarized and an authorization to proceed to launch was recommended. The assembled multitude was then asked to speak up "if anyone here wishes to take issue with this recommendation." I could feel the heat on me, as did everyone else in the room; there were three problems that had not been completely resolved, and the launch was on the line. The likelihood of them causing a problem was small, but a failed launch would be a disaster; there was no backup XTE. What to do? I did not raise my hand nor did anyone else. We were all thus co-opted into the decision to launch.

My hesitation was based on experience in our rocket program. Almost any launch will have some minor residual problems that cannot be completely resolved. If launches were delayed until each and every (minor) concern were completely resolved beyond any doubt, one might never launch.

On the other hand, if one accepted this philosophy and a failure results, one would regret taking that position. On balance, it is a judgment call based on the perceived level of risk. The danger is that one's perception of risk can change with time. This was an important causative effect in the tragic losses of two Space Shuttles and their crews.

Attempts 6 and 7: success, December 30, 1995

With a go-ahead authorized, another attempt, the sixth, took place on December 29. It too was scrubbed due to high-altitude winds. Finally, on December 30, 1995, on the seventh attempt, the XTE was successfully launched into orbit at 8:48 a.m. (Fig. 20). I did not see the liftoff because I was listening to the reports on the voice circuits in the VIP room. However, I could feel the engine vibrations through the walls, and I did walk outside to see the bird shortly after liftoff.

By pre-arrangement, the satellite was named the *Rossi X-ray Timing Explorer* after Bruno Rossi, the founder of MIT's cosmic ray group, who had died in 1993. Under his leadership, the group began pioneering research in X-ray astronomy and interplanetary plasma research, both of which must be carried

Figure 20: Liftoff, Dec. 30, 1995

Launch of the X-ray Timing Explorer at Kennedy Space Center, December 30, 1995, at 8:48 am (ET). Note the just-released umbilical cords. [KSC/NASA].

out from space. It was Rick Rothschild who had proposed the Rossi name, which was gratifying to us at MIT who had worked with and under Rossi. It would have been considered presumptuous for us to propose it. This was then proposed to Jean Swank and her colleagues and to Alan Bunner of NASA Headquarters, and all agreed it was a good plan. We also checked with Rossi's family making sure they understood that, if things went awry, the Rossi name could become a subject of ridicule or scorn, as had happened, temporarily, with the *Hubble* mission in 1990.

I was deputized to check with the Italians who were about to launch their own X-ray astronomy satellite and could have been thinking of naming it after Rossi whose memory was revered by the Italian scientific community. I called Livio Scarsi, a longtime friend and the prime mover of their program. He was pleased to hear our plan, because as he said, "This frees us to name our satellite after Giuseppe Occhialini," who was another famous Italian physicist, two years Rossi's junior and Rossi's first PhD student. The Italian satellite, named *Beppo-SAX"* was launched successfully four months after XTE on April 30. *Beppo* is the nickname for Giuseppe, and "SAX" (*Satellite per Astronomia a raggi X)* was the original acronym of the satellite.

Thus on December 30, 1995, the XTE became the *Rossi X- ray Timing Explorer* (RXTE). Regretfully in retrospect, we neglected to encourage our colleagues to refer to it simply as the *Rossi*, rather than our usual "RXTE." Fred Lamb used *Rossi* himself in verbal conversation, but it went no further.

ASM crisis after launch

About a week after launch, we turned on the ASM, and initially it performed as expected. But over the next several days, its data stream exhibited irregular strange bursts of counts. We eventually came to realize these were high-voltage breakdowns (electrical arcing). The problem worsened, and we began to see *continuous* breakdown in two of the ASM detectors. Each detector had eight primary anodes so the loss of one of them, or even two, was tolerable. However, the electrical noise of such a breakdown swamped the signal on the other working anodes in that detector so the entire detector was useless. Each of the three ASM units had a single detector, so two thirds of our capability were seemingly useless!

This breakdown occurred in the two detectors that had no pinhole leak and thus had *not* experienced oxygen poisoning! We spent about three weeks analyzing data trying to understand the problem. During much of this time, we had shut down the offending counters by turning off the high-voltage power to them, to limit any damage the breakdown might be causing.

Those were desperately unhappy times as they threatened to render our entire ASM instrument only minimally effective. This partial blinding of the ASM would greatly reduce the science it might accomplish. It would also limit the effectiveness of the PCA and HEXTE instruments in acquiring targets of interest. It would also sully MIT's reputation in space science and our reputations within and outside MIT. As principal investigator, I would surely take a lot of the heat and pity. Those were dark, discouraging days indeed. Garrett Jernigan recently told me he still recalls the look on my face as we faced this issue, one of utter anguish.

During the first in-orbit breakdown events, which at first were quite puzzling, we had a meeting at Goddard with scientists from the Goddard PCA group; they were proportional counter experts. There, just as the meeting was breaking up with no clear conclusion on how to proceed, Pete Serlemitsos, a superb instrumentalist at Goddard, commented under his breath, "I cannot understand how anyone would build and fly a proportional counter without adjustable high voltage."

This was quite a damning statement, which I took to heart, because taking such precautions was part of being a good scientist. Our detectors ran at a predetermined fixed voltage, which was either on or off. Our engineering group had a philosophy of keeping systems as fundamentally simple as possible, because every additional control or complication is another potential failure point. With this approach, we had had a string of highly successful flight instruments. Peter had a point, but his way wasn't the only way. Still, his comment certainly did not lighten my mood.

As we were considering our options, I recalled seeing how a breakdown had stripped part of an anode bare of its carbon coating at Metorex, and either Al Levine or I suggested that we let the breakdown proceed by turning on the high voltage and leaving it on. Perhaps, after the arcing had worked its way down the entire length and burned all the carbon off that anode, the breakdown would cease. Then, with the cessation of the electrical noise, we could use the undamaged wires in the detector. We tried it on a prototype counter in the laboratory, and it recovered after a couple days as we hoped.

So, we turned on the voltage in one of the defective flight units in the orbiting RXTE and let the breakdown continue for one day and then for another whole day. It seemed as if it would never stop. Then, finally, on the third day or perhaps later, the breakdown sputtered to a stop, and the other wires performed flawlessly. We repeated this for the other arcing detector successfully, and again perhaps one or two times later when new breakdowns occurred. Finally, we had three stable counters that performed as designed, albeit with a few missing anodes, and thus a modest loss of sensitivity. They continued to provide data for the next 16 years. It seemed a miracle.

What could we have done better? In retrospect, adding a bit more quench gas to the detectors or lowering the high voltage would have solved the breakdown problems we were having before launch. The hoped-for sensitivity of our detectors argued for doing neither, but in retrospect that was a mistaken judgment. If the high voltage had been adjustable in flight, lowering it would have been an option.

We at MIT were so pleased to finally have a complete well-functioning ASM in orbit that we gathered around the engineering-unit ASM for an impromptu photo opportunity (Fig. 21).

Looking back on it now, I realize that, even if we had permanently lost two of our three cameras, the third was still a powerful instrument with a lifetime of years, if not the entire mission, given the slow gas leak. It was one of the two azimuthal cameras, which meant that it would have surveyed about 70% of the sky with each 90-minute scan, determining source intensities, albeit with one-dimensional lines of position. A small maneuver of the spacecraft could provide a second crossing line if the source position was previously unknown. An outpouring of excellent ASM results would soon mask our initial troubles, as we have seen with other missions, e.g., *Hubble* and *Chandra*.

Strangely, I have no recollection of seeing this more positive spin at the time. That seems so unlike me; I have always been prone, when confronted with a seemingly impassable obstacle, to quickly think of a worst-case but livable strategy before settling down to finding the optimum strategy.

Figure 21: MIT RXTE team with ASM
All-Sky Monitor engineering unit with our science team, after launch. From the left: Hale Bradt, Edward Morgan (peeking), Ron Remillard (below), and Alan Levine. [Arlyn Hertz]

VII
Science Highlights

WAS THE SCIENTIFIC PAYOFF WORTH worth
all the effort? There was no certainty that this would be the
case. The mission followed Bruno Rossi's dictum that it pays
to explore new domains of physical phenomena. Unexpected
discoveries could result, and they did for RXTE. Was there
an absolutely mind-bending, headline-grabbing, Earth-
shaking, Nobel-winning discovery?[7] At present, it seems
there was not. But there were important new surprises that
were impressive to the scientific community and that led to
major advances in our understanding of neutron stars, black
holes, and more.

More than 2100 papers with RXTE data were published
in refereed journals (through 2012) with an average of 22 citations, and there were 60 high-impact papers that had more than
100 citations. Ninety PhD theses were written based at least
in part on RXTE data. Another mark of its science yield was
that the High Energy Astrophysics Division of the American

Astronomical Society (HEAD/AAS) awarded its Bruno Rossi Prize on four different occasions for RXTE science.

After its initial two years of operations, RXTE faced biannual Senior Reviews wherein it went head-to-head with other operating NASA astrophysics missions seeking funds for another two years of science operations. The science merits of the several missions were judged by a committee of scientists chosen from a wide range of astrophysics disciplines. Seven successive committees judged favorably the uniqueness of RXTE capabilities and the quality of its scientific output. This led to NASA funding for continuing satellite operations and scientific data analysis for a total of 14 additional years.

RXTE science operations finally ceased on January 3, 2012, and the satellite was deactivated two days later, after a full 16 years of observations. Thereafter it silently continued to orbit the Earth until its entry into the atmosphere on April 30, 2018. During its 22-year journey, it orbited the Earth some 120,000 times. That distance would have taken it to the sun and back some 15 times or instead out to the orbit of Neptune once —it's a long way out there.

RXTE observations made by scientists at institutions in the US and elsewhere covered a wide variety of scientific topics from white dwarfs, stellar black holes, and neutron stars to massive black holes at the centers of galaxies. Jets, spins, magnetic fields, pulsing mechanisms, and evolution were all investigated. The 100% guest program stimulated a huge range of investigations by more than 400 guest observers. Many RXTE observations were carried out simultaneously with studies by other observatories, both ground-based and orbiting, at radio to TeV gamma-ray wavelengths. It was often an RXTE discovery that triggered observations by another observatory.

Our group at MIT was involved in much of this but was only a small part of the overall effort. There were huge contributions, both observational and theoretical, from others at MIT and elsewhere in the US (e.g., scientists at Goddard, and UCSD, Fred Lamb at U. Illinois, and Jeff McClintock and Ramesh Narayan at Harvard/Smithsonian), and also internationally, from The Netherlands (e.g., Michiel Van Der Klis), Italy (e.g., Tomaso Belloni and Luigi Stella), Japan (e.g., Yasuo Tanaka), and the UK (e.g., Rob Fender and Andrew Fabian). At MIT, in addition to our team, users of RXTE included Saul Rappaport, Walter Lewin, Deepto Chakrabarty. and George Clark.

The day-to-day management of the science observations of the orbiting RXTE was carried out by the Science Operations Center at Goddard directed by X-ray astronomer Frank Marshall. He and project scientist Jean Swank heroically stayed on top of the entire effort for all those 16 years, pulling together the bi-annual Senior Review reports, managing the guest-observer program, and deftly reconciling competing interests. The huge amounts of RXTE data that continuously flowed into Goddard were organized and made available to observers by Goddard's HEASARC.

My own role in extracting science from RXTE data was not particularly strong. Younger colleagues were much more active than I. They had the energy, the knowledge for complex analyses of data, and the imagination for new types of investigations. I worked closely, though, with three of our students on RXTE results: Bob Shirey on studies of the neutron-star binary Circinus X-1, Linqing Wen on studies of the black hole candidate Cyg X-1 and searches for periodicities in the ASM data, and Don Smith on studies of gamma-ray bursts in the ASM data. Our other two graduate students at this time (Michael

Muno and Dacheng Lin) worked with Ron Remillard, Deepto Chakrabarty, and Ed Morgan on X-ray timing and spectral studies of accreting black holes and neutron stars.

I enjoyed continuing discussions with my ASM/EDS team: Ed Morgan, our fast-timing expert who facilitated and participated in many of the exciting discoveries of variability on millisecond time scales; Al Levine who worked on and reported many results from the ASM monitoring of source intensities and from PCA pointed observations pertaining to compact-star binary orbits; and Ron Remillard who carried out observations, both X-ray and optical, of stellar black-hole systems. In addition, these three scientists monitored and maintained the in-flight operations of the EDS and ASM for the entire 16-year mission.

The Rossi Prizes

The citations of the four Rossi prizes describe highlights of RXTE science. The prizes honor the recipients "for a significant contribution to High Energy Astrophysics...."

> The 1999 Rossi Prize of the High Energy Astrophysics Division of the American Astronomical Society is awarded to Drs. Jean Swank and Hale Bradt for their key roles in the development of the *Rossi X-Ray Timing Explorer*, and for the resulting important discoveries related to high time resolution observations of compact astrophysical objects.

> The 2003 Rossi Prize of the High Energy Astrophysics Division of the American Astronomical Society is awarded to Robert Duncan and Christopher Thompson for their prediction, and to Chryssa Kouveliotou for her observational confirmation, of the existence of magnetars: neutron stars with extraordinarily strong magnetic fields.

> The 2006 Rossi Prize of the High Energy Astrophysics Division of the American Astronomical Society is awarded to Tod Strohmayer, Deepto Chakrabarty, and Rudy Wijnands for their pioneering research that revealed millisecond spin periods and established the powerful diagnostic tool of kilohertz intensity oscillations in accreting neutron star binary systems.[8]

> The 2009 Rossi Prize of the High Energy Astrophysics Division of the American Astronomical Society is awarded to Charles D. Bailyn, Jeffrey E. McClintock, and Ronald A. Remillard for their measurement of the masses of Galactic black holes.

The 1999 award recognizes Jean and me for our contributions to the development of RXTE but fails to recognize others who played equally important roles, e.g., Steve Holt, Charlie Pellerin, Rick Rothschild, and Fred Lamb. Not to be overlooked were the efforts of the high-energy-astronomy team at NASA Headquarters, Albert Opp, and later Alan Bunner and Lou Kaluzienski who kept pushing for the mission. The science part of our award refers primarily, I would think, to the spectacular kilohertz discoveries described below and explicitly recognized with the 2006 prize. I was not directly involved in that work, whereas Jean worked with Tod Strohmayer in his investigations. In fact, all who participated in the development of RXTE deserve a share of the credit for its important discoveries.

I now describe some of the major RXTE results that attracted wide attention. They demonstrate well the power of the RXTE instruments in exploring the X-ray temporal variability of the cosmos. Each of these results was considered quite spectacular by the astronomical community. These selected results are my choices. Others would surely have chosen and emphasized differently.

To set the stage for these examples, I first introduce the concept of an X-ray binary star system for those not familiar with it.

Accreting X-ray binary star systems

An "X-ray binary" is a two-star system wherein one star is a normal star, that is, a hot ball of gas like our sun, and the other is a compact object such as a black hole or a neutron star. The two stars revolve about their common center of mass, as they must to keep from falling into each other. One should appreciate the huge disparity of sizes, the compact object being the size of Manhattan and the normal star being the size, perhaps, of 100 Earth diameters or more, a size disparity of order 100,000. Think of a football field as the normal star and a 1-mm ball-bearing as the neutron star. Yet their masses can be comparable! The compact object is incredibly dense.

For a black hole we interpret its "size" as that of the event horizon from within which no information can escape. We refer here to *stellar black holes* with masses comparable to and somewhat greater than the sun and sizes a few times larger than a neutron star.

If the outer layers of the normal star come sufficiently close to the compact object, its gaseous outer layers can flow down toward the compact object forming an *accretion disk* (Fig. 22); think of Saturn's rings. If there is an energy-loss mechanism in the accretion disk, for example viscosity (friction), the gas will spiral inward and eventually down onto the surface of the neutron star, or into the black hole.

As it does, the intense gravitation field causes it to dramatically lose potential energy, which appears as a huge gain in kinetic energy. When the gas flow is randomized in the disk or on the surface, its temperature rises to ten million degrees or more. At such temperatures, X-rays are the dominant component of the radiant energy, and this is what X-ray observatories such as RXTE can study. Figure 22 is an artist's conception of such a binary system.

In the course of studying these systems, of which there were some 100 known in our Galaxy, X-ray astronomers found that certain neutron-star systems were *pulsars* that emit pulses of radiation at regular intervals. The origin of these is believed to be a hot spot of X-radiation on the neutron-star surface at the magnetic pole where the accreting material streams down onto the surface. As the neutron star rotates about its spin axis, the hot spot disappears and reappears every rotation, giving regular pulses of X-rays that directly indicate the spin period of the neutron star. These objects are often called *accretion-powered pulsars*.

Kilohertz oscillations discovered

RXTE was hugely successful in finding kilohertz oscillations (periodic intensity changes on millisecond time scales) in both neutron-star and black-hole binary systems. Both types of systems have been reviewed respectively by Michiel van der Klis[9] and by Ronald Remillard and Jeffrey McClintock.[10] These detections were a major breakthrough for RXTE science.

The first detections occurred in mid-February 1996, a mere seven weeks after RXTE was launched. Searches for the long-sought kilohertz oscillations were carried out by two teams, one led by Michiel van der Klis at the University of Amsterdam and the other by Tod Strohmayer at Goddard, and both hit pay dirt within days of each other. The former team found kilohertz oscillations in Sco X-1—the first known X-ray source discovered back in 1962—and the latter found them in a source known as 4U 1728–34. (The 4U refers to the catalog listing the source and the digits refer to the location in the sky.)[11]

Figure 22: X-ray binary, artist's conception
Artist's conception of an X-ray binary system showing the large normal star to the left, which is losing gas through an accretion disk to the compact object, a black hole or neutron star at the white spot at the center of the disk, but the compact star is about 1000 times smaller than the white spot. The disk of accreting gases surrounding the compact object is hottest in its inner parts, which therefore emit in the X-ray band. In many black-hole systems, jets of gaseous material are ejected up and down along the rotation axis, as illustrated here. [NASA/Dana Berry]

The two teams reported their results in back-to-back IAU Circulars, which are used for rapid notification of astronomical events to fellow astronomers, but also to get credit for being first to detect (discover) something new and important. Both teams later published refereed papers in the same issue of the prestigious *Astrophysical Journal Letters*.[12, 13] Ed Morgan and Walter Lewin of MIT were co-authors on the van der Klis paper.

This discovery was a huge deal for all of us on the RXTE team; kilohertz variability had long been an object of our efforts and here it was clear and strong in two different sources. It boded well for RXTE's future. We were vindicated and looking forward to exploiting these discoveries with future studies and searches. Eventually, three types of oscillations were found in neutron-star systems and one type in black-hole systems.

Kilohertz oscillations in neutron-star binaries

Burst oscillations (nuclear powered pulsations)

In a neutron-star binary system, the matter streaming down onto the neutron star—largely hydrogen—may fuse ("burn") into helium either during the infall or on the surface. The helium accumulates on the surface until it reaches sufficient depth, and hence pressure at its base, to suddenly start fusing into even heavier elements such as carbon, oxygen, and even to iron and beyond. It does so explosively, like an H-bomb, releasing a huge burst of X-rays. As accretion continues, such bursts continue to occur at intervals of hours to days. The discovery and the interpretation of these nuclear-powered "X-ray bursts" pre-dated the launch of RXTE.

The X-ray source, 4U 1728–34 is such a burst source. The February RXTE observations of it by the Goddard team immediately yielded kilohertz oscillations in both the burst flux but also in the persistent (non-burst) flux. The two types were markedly different in character. This was a double win for RXTE!

The burst oscillations were discovered in a burst lasting about 20s. During this period, the X-rays were found to be pulsing at 3–10% of the flux with pulse intervals of about 3 ms (milliseconds), or equivalently at frequencies of 361–364 hertz (Hz or cycles per second)! Or restated, the X-ray intensity rapidly modulated at those frequencies, which for sound waves is about the F# above middle C on the musical scale. Even though the frequencies are less than one kilohertz, namely 0.36 kilohertz,

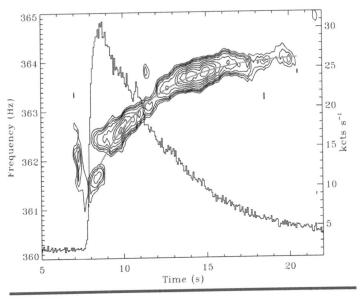

Figure 23: X-ray burst with kilohertz oscillations

The intensity and frequency drift of an X-ray burst from the low-mass X-ray accreting binary system, 4U 1728–34. The histogram gives the intensity as a function of time in kilocounts per second (right axis). The counts are binned in time intervals (bins) of 62.5 ms. The burst reaches 30,000 counts per second at its peak and lasts about 20 s. The steady flux of X-rays in the first few seconds is due to gaseous matter from the normal star spilling over onto its neutron-star companion. The sudden rise at time 8 s is the beginning of an X-ray burst due to accumulated accreted material on the surface of the neutron star erupting in a thermonuclear (as in an H-bomb) explosion. The contours show the strength of oscillations in the X-ray signal together with its frequencies (left axis), again as a function of time. [Strohmayer, pvt. comm.; see Strohmayer et al. ApJ 469, L9 (1996)]

they are considered to be in the domain of kilohertz variability. The ringing lasted only about as long as the 20-s X-ray burst. These oscillations could be called *nuclear-powered pulsations* to distinguish them from other types.

Figure 23 shows an X-ray burst from 4U 1728–34 containing oscillations. The histogram gives the count rate of detected

X-rays (kilocounts per second, right axis) as a function of time in seconds. Compare the huge intensity of the peak (30,000 counts per second) to the much lower persistent emission (1500 counts per second) just before the eruption, a factor of 20. And note the burst duration of about 20 seconds. That is one dramatic blast of X-rays!

The oscillation frequencies (left axis) of the burst are given by the contours in the figure as a function of time. The height of the contours gives the strength of the oscillations. The oscillations are so fast, with a period only 3 ms, they can not be seen in the histogram of X-ray intensity, which has time bins of 62.5 ms.[14] A strong ringing starts at about 362.2 Hz, decreases to 361.5 Hz and drifts upward during the burst to an asymptotic value of about 364.2 Hz. The statistical precision of these data show the power of the PCA instrument, namely its ability to collect large numbers of X-rays in short time periods.

These data and others obtained later from this and other similar sources led to a partial explanation of the pulsations, namely that the asymptotic frequency, 364 Hz, is the spin frequency of the neutron star. The oscillations would be due to a hot spot of burning hydrogen on the surface coming into and out of view as the star rotates. The lower frequencies before reaching 364 Hz could be due to the burning surface layers being elevated tens of meters by the explosion and lagging to conserve angular momentum. As they fall back to the surface, the observed frequency rises to that of the neutron-star spin.

This observation thus reveals a neutron star that rotates around its axis at the amazing rate of 364 times a second. Think of a spherical Manhattan spinning that fast.

Such rapidly rotating neutron stars had been known to emit radio waves; they had no stellar companion and were known as *millisecond pulsars* (discussed below). But no X-ray emitters had ever been known to exhibit such high spin rates.

RXTE had found the sought-after kilohertz oscillations. Moreover, they were periodic and also occurring during the drama of a huge thermonuclear outburst on a neutron star. Both characteristics made them powerful diagnostic tools. We were all overjoyed to find this new method for probing neutron-star physics.

Quasi-periodic oscillations (QPO)

The early (February 1996) excitement was further enhanced by the simultaneous detection of *quasi-periodic oscillations* (QPO) at kilohertz frequencies in the persistent (non-burst) flux from both 4U 1728–34 by the Goddard team and in Sco X-1 by the van der Klis team. (Sco X-1 is not a known burst source.)

The "quasi" indicates that the detected frequency of oscillation is not fixed; it can change in time by large amounts, more than can be explained by a simple Doppler shift or gases rising during an X-ray burst. During a single observation, the detected frequencies will be more-or less fixed, but on another day, they be might found at a much higher or lower frequency or even be nonexistent. Figure 24 shows the distribution of power as a function of the frequency of oscillation in Sco X-1, for about three hours of data. The peaks indicate that much of the power is concentrated near the two well-separated oscillation frequencies, 600 and 850 Hz.

With repeated observations of the numerous sources exhibiting this phenomenon, the variation of the dominant frequencies can be studied as a function of changing luminosity (accretion rate) and spectral hardness (X-ray color) to place severe constraints on models. The oscillations must stem from regions that can give rise to such high frequencies, namely the

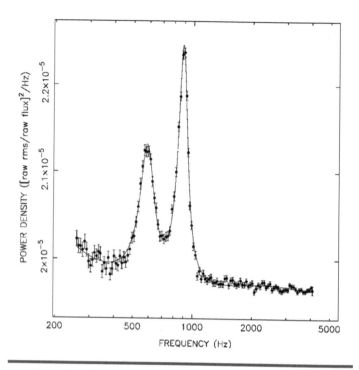

Figure 24: Quasi-periodic oscillations in Sco X-1
Power density spectrum of Sco X-1 containing about three hours of data. This plot shows the distribution of X-ray power as a function of the oscillation frequency of the X-ray intensity. Peaks of power with very high statistical significance are evident at about 600 Hz and 850 Hz. [van der Klis, et al., ApJ 481, L97 (1997)]

vicinity of the spinning neutron star with its strong co-rotating magnetic field and rapidly orbiting accreting material.

MILLISECOND ACCRETION-POWERED PULSARS)

In 1982, radio astronomers made a substantial advance with the discovery of a *rotation-powered pulsar* spinning at 642 Hz. This was hugely fast; its spin period was only 1.6 ms. It and similar examples discovered later were called *millisecond pulsars.*

They were all isolated neutron stars, with no binary companion, and were all radio emitters. The power for the radiation arose from their rotational kinetic energy; their spins were thereby very gradually slowing. Hence the name, rotation-powered pulsar. They were fast-spinning versions of the long-known, since 1967, neutron-star, rotation-powered radio pulsars, which had much slower spin rates, typically about 1 Hz.

The radiation from these pulsars emerges, it is thought, from the magnetic pole of the neutron star as a beam of radio waves. If the magnetic pole is offset from the spin axis, each rotation of the star will produce a flash of radio waves at places the beam transits, like a lighthouse. If the beam transits Earth, our radio astronomers will detect a pulse of radio waves for each rotation of the neutron star. The pulse intervals directly represent the spin period of the neutron star, which, in our examples here would be 1.6 ms (642 Hz) and 1 s (1 Hz) .

Evidence indicated these millisecond pulsars were in a late phase of their lives and had once been slow rotation-powered radio pulsars of the kind long known. If so, what were the torques that spun them up to kilohertz frequencies?

The favorite theory was that they were spun up by accretion torques in an accreting binary system. The infalling gases in the accretion disk spiral around the neutron star at angular velocities greater than that of the star's spin and thus impart an accelerating torque as they settle down onto its surface. This continues until the neutron-star spin reaches some equilibrium spin rate at the observed kilohertz frequencies.

But the radio-emitting millisecond pulsars have no normal star companion to provide the accreting gas; they are isolated stars. The hypothesis was that the pulsar (a neutron star) was *formerly* one member of an accreting binary system, and the strong X-radiation during the accreting phase had completely vaporized the donor star, leaving behind only an isolated rapidly spinning neutron star. Such a neutron star is sometimes called a "black widow"; it devours its mate.

Now, since accretion onto a neutron star yields copious X-ray emission, X-ray astronomers should be able to detect such a system—a "millisecond accretion-powered pulsar," and RXTE was ideally suited to make such a search. In fact, finding such a pulsar was considered by some to be the Holy Grail of RXTE's search for kilohertz oscillations. An accreting millisecond pulsar was a phenomenon for which the evidence was compelling; it should be there for RXTE observers to find.

Well, it wasn't so simple. It wasn't until April 13, 1998, more than two years after the RXTE launch and many studies of kilohertz oscillations in numerous sources, that success was achieved. Observation of a source known as SAX J1808.3–3658 by the van der Klis team revealed strong pulsations at the very steady frequency of 401 Hz. The discovery was reported in two papers in *Nature*, the first by Rudy Wijnands and Michiel van der Klis of the University of Amsterdam announcing the discovery, and the second by MIT's Deepto Chakrabarty and Edward Morgan who used the Doppler shifts of the pulsing to map the orbit of the accreting neutron star (Fig. 25).

The hypothesized accreting phase of a millisecond radio pulsar was caught in the act by RXTE! The Holy Grail had been found.

The long-term steadiness of the 401-Hz frequency demonstrated without question that it represents the spin frequency of the accreting neutron star. (The huge inertia of a neutron star keeps it spin rate nearly constant.)

The data from this source mapped precisely the orbit of the neutron star in its binary system. (Both stars rotate about their

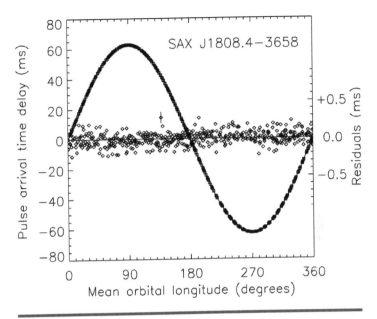

Figure 25: Accreting millisecond pulsar orbit

Time delays from the SAX J1808.4–3658 binary system of the 401-Hz pulses as a function of the orbital longitude of the binary (think "time" where 360° corresponds to the 2.01-hr orbital period). As the neutron star orbits the binary center of mass, it moves cyclically toward and away from the Earth observer. At its greatest distance, the pulses received at Earth have to travel a bit farther and hence arrive slightly delayed. Similarly they are advanced when the neutron star is closest to Earth. The sinusoidal delay curve shown here thus plots out the projected orbit—with radius 62.809 light-milliseconds—of the neutron star with great precision. The residuals to the best fit are only about 0.1 ms (points about horizontal axis and scale on right). The pulsing in this source was discovered in RXTE data by Rudy Wijnands and Michiel van der Klis [Nature 394, 344 (1998)]. The Doppler orbit shown here was reported by Deepto Chakrabarty and Edward Morgan in the following paper [Nature 394, 346 (1998)].

common center of mass.) The pulsing of the neutron star acts like a ticking clock. Each clock tick travels to us as an enhanced rate of X-rays, which have to travel farther when the neutron star is on the far side of its companion, the normal star, and less far when it is on the nearer side. By plotting the delays (plus and minus) in arrival times of the pulses (Fig. 25), the projected orbit of the neutron star in the X-ray binary can be mapped to great precision.

The three types of kilohertz oscillations in neutron-star binaries (burst, quasi-periodic, and accreting) established the richness of this phenomena as a diagnostic tool. Each had its own distinctive characteristics, and each was found in numerous sources. RXTE continued studies of these systems for the rest of its 16 years in orbit.

Kilohertz oscillations in black-hole binaries

Mapping the orbits of the stars in an accreting X-ray binary system provides information about the masses of the normal star and the compact object emitting the X-rays. (This is usually done by measuring the Doppler shifts of the optical spectral lines from the normal star.) Some systems are found to have compact objects with masses exceeding that deemed possible for a neutron star, more than about three solar masses. They are thus presumed to have a black hole as the compact partner of the binary system. RXTE has studied the frequency distribution of the time variations of the energy flux in X-rays from several such systems. Six examples are shown for five sources in Fig. 26. In each, there is a peak of power at frequencies that range up to 300 Hz. The frequencies can change with time as seen for the two plots for source XTE J1550–564.

These oscillations arise from the hot X-ray emitting gases in the near vicinity of the black hole and hence provide diagnostics of those regions. They could be hot spots in the gases orbiting the black hole or oscillations of the accretion disk. Again, the kilohertz capability of RXTE opened new territory for theoretical explanations of the relativistic behaviour of plasmas in strong gravitational fields.

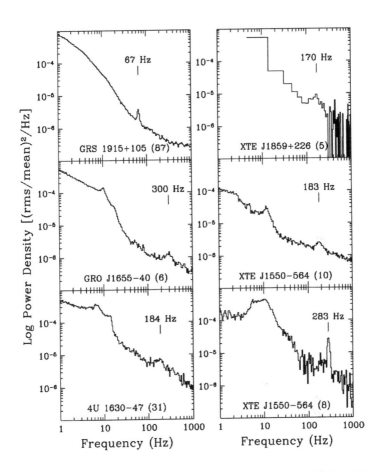

Figure 26: Kilohertz oscillations in black-hole binaries

Power density spectra for five black-hole binary systems with two separate observations for J1550–564. Each plot shows how much power there is at each frequency of intensity oscillations from 1 to 1000 Hz. Each of these systems consists of a normal star spilling matter onto (or into) a stellar black hole. The radiation comes from the gaseous material immediately surrounding the event horizon of the black hole. The indicated peaks represent ringing (intensity changes) at the indicated frequencies. [R. Remillard, in "Evolution of Binary and Multiple Stars," ASP Conf. Series, v. 229, eds. P. Podsiadlowski, et al., p. 503; see also R. Remillard and J. McClintock, ARAA 44, 49 (2006)]

X-ray variability over months and years

The All-Sky Monitor data were, of course, my group's favorite because we had developed the ASM instrument. Over time, the results from it were spectacular. The assembly of three cameras on one rotatable platform advanced with six-degree steps. It stopped at each step for a 90-second exposure to record "images" of three 10° × 100° slices of the sky. Most of the sky would thus be imaged during each 90-minute orbit of RXTE. These exposures, taken hour after hour, day after day, and year after year, yielded a wealth of intensity points of the brightest 100+ X-ray sources. From these, we were able to construct "light curves" (intensity vs. time) for these sources over the entire 16-year period of the mission. Figures 27 and 28 show the light curves obtained with the ASM over a period of five years for a sampling of sources with interesting behaviors.

Figure 27 contains the five-year light curves of seven "persistent" X-ray sources. One of these (Mk 421; lowest plot) is the active core of a distant galaxy, known as an *active galactic nucleus*, and two (GX 339–4 and LMC X-3) are black-hole binary systems. The other four are likely neutron-star binaries. These sources exhibit a wide variety of variability: periodic (SMC X-1 and Her X-1), aperiodic (LMC X-3, Cyg X-2), a year-long bright state (GX 339–4), and sustained "off" (undetectable) states (3A1942+274 and Her X-1).

Figure 28 shows the five-year light curves for seven transient or recurrent X-ray sources. A source will be undetectable for months or years and then suddenly appear, seemingly from nowhere, for periods of days to months, after which it disappears again. The X-ray flux is a direct indicator of accretion taking place. Hence these light curves are direct evidence of episodic gaseous outflow from the donor star or to changes

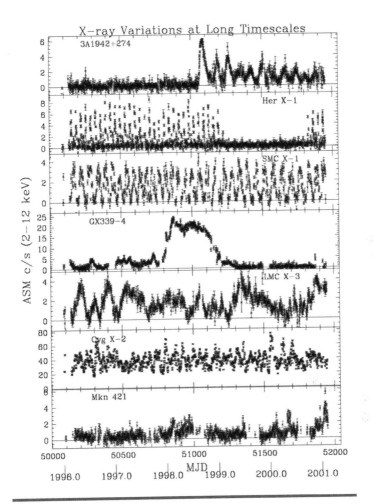

Figure 27: Five-year X-ray light curves, persistent sources

Five-year light curves (intensity versus time) of seven "persistent" X-ray sources from the All-Sky Monitor on RXTE. Measured counts (X-rays) per second are plotted as a function of time in years or days (modified Julian Day, MJD). One of these (Mk 421; lowest plot) is the active core of a distant galaxy, known as an active galactic nucleus, and two (GX 339–4 and LMC X-3) are black-hole binary systems. The other four are likely neutron-star binaries. Two (LMC X-3 and SMC X-1) are in the nearby Magellanic Clouds; the others, excluding Mkn 421, are in the Milky Way Galaxy. [Courtesy of Ronald Remillard and Alan Levine; see also R. Remillard and J. McClintock, ARAA 44, 49 (2006)]

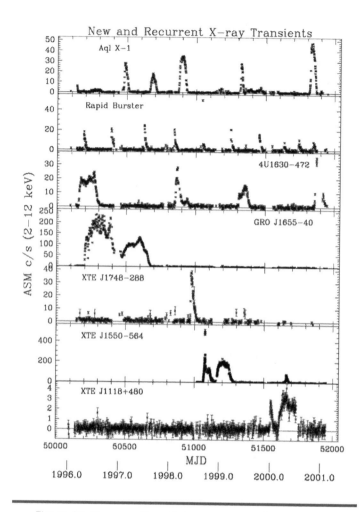

Figure 28: Five-year X-ray light curves, transients

Five-year light curves for seven transient or recurrent X-ray sources from the ASM on RXTE as a function of time (modified Julian Day, MJD). The top two sources are likely neutron-star binaries, and the lower five are black-hole binaries, either confirmed or likely. Such RXTE observations led to optical measurements that yielded the masses of several black holes. [Courtesy of Ronald Remillard and Alan Levine; see also R. Remillard and J. McClintock, ARAA 44, 49 (2006)] of state in the accretion disk that cause it to stop or start dissipating energy.

The ASM data alerted astronomers worldwide to newly detected outbursts and sometimes to previously unknown systems. Observations of such transients led to optical measurements that yielded the masses of several black holes. They show dramatically the value of having an ASM to guide observations.

The extreme variability of the X-ray sky shown in these plots is markedly different than the visible sky. Similar plots of the intensity of most visible stars in optical light would be absolutely flat lines to very high precision. The drastic variability of the X-ray sky is also demonstrated in an animation produced from RXTE/ASM data. It shows the changing brightness of the brighter X-ray sources in the sky over four years (1996–99) of the mission. The 7-minute video is available on youTube.com (search "rxte asm sky").

In general, the ASM was successful in its stated objectives, namely, to facilitate new discoveries—often with follow-up PCA or optical/radio observations—and to provide observers the timing and context for their PCA observations. Results directly from the ASM data included new black-hole candidates, periodicities, positions of new transients, and locations of gamma-ray bursts. It was humankind's first sustained bird's eye view of variability over the *entire* X-ray sky.

Cyclotron lines with the HEXTE

The HEXTE experiment produced a number of solid results also. One of its objectives was to probe X-ray sources for evidence of strong magnetic fields though the detection of *cyclotron lines*. One example of its success is the *energy spectrum*

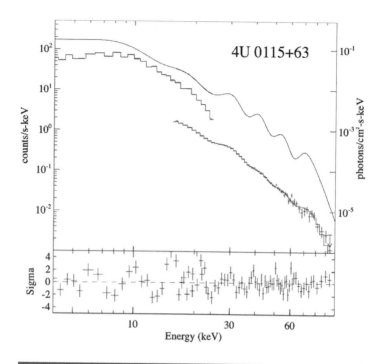

Figure 29: Energy spectrum with five cyclotron lines

*Energy spectra of the X-ray source 4U 0115+63 obtained with the PCA (low energies, left histogram) and HEXTE (high energies, right histogram). The histograms are measured **count-rate** spectra with fits (left scale) plotted as a function of the energy of individual X-rays. The upper smooth curve is the combined best-fit **incident photon** spectrum (right scale); it exhibits five cyclotron lines. The lowest plot shows how well the fits match the count-rate data. [Courtesy, R. Rothschild; see Heindl et al., BAAS, 32, 1230 (2000)]*

of the X-ray source 4U 0115+63 shown in Fig. 29, where several quantities are plotted against the X-ray energy in units of keV. The two stepwise truncated curves show the detected *count rates* from the PCA (lower energies) and HEXTE (higher energies). The upper smooth curve takes into account the efficiency of the detectors, which depends on X-ray energy, to give the best-fit combined *incident* X-ray spectrum. It exhibits five peaks, which are interpreted as cyclotron lines. This was the first time more than two such lines in any source had been found.

These data allow theorists to model the emitting system to obtain geometries and an estimate of the magnetic field, in this case about 10^8 tesla (T). It had long been thought that neutron stars emitting pulsed radiation had high magnetic fields of such magnitude, but cyclotron studies provide direct confirmation. It should be noted here that "magnetars" (discovered with RXTE) are neutron stars with fields as strong as 10^{10} T. (The Earth's magnetic field at the surface is less than 1/10,000 T and the strongest MRI magnets produce fields of about 3 T.)

Microquasar GRS 1915+105

Observers have used RXTE to study in great detail the nature of a source known as GRS 1915+105. It is highly variable in X-rays and exhibits flares in the radio, infrared, and X-ray bands. Radio astronomers have observed jets of plasma (ionized gas) emerging from the object at relativistic speeds—close to the speed of light. With strong likelihood, this system is a black-hole binary with a black-hole mass of 10–15 solar masses. The object, with its jets, is reminiscent of *extragalactic quasars*, which are at great distances and powered by massive black holes with masses of 10 million solar masses or more. Nearby black-hole systems, such as GRS 1915+105 are much less powerful and hence are sometimes called *microquasars*.

Study of microquasars can give valuable insight into the behavior of extragalactic quasars. Their much closer distances give greater fluxes at Earth despite their lesser luminosities. Moreover, the time scales of matter motions near the black hole, e.g., the orbital period, scale as the mass of the black hole, and these matter motions can give rise to X-ray intensity variations. Thus, a 10-minute variation in a microquasar of 10 solar masses (like GRS 1915+105) would last 10 million minutes (19 years) in a 10-million solar mass extragalactic quasar. Thus, the study of intensity variations in microquasars over hours can help us understand the processes in extragalactic quasars extending over centuries.

The GRS 1915+105 source showed extreme X-ray variations over a four-year time scale in the RXTE ASM data and also in radio emissions (Fig. 30) and a bizarre set of repetitive X-ray intensity modes when viewed by the PCA over different 50-minute periods (Fig. 31). The X-ray observations represent activity at the innermost regions of the accretion disk.

Simultaneous X-ray, radio, and infrared observations of this object (Fig. 32) appear to catch the injection of matter from the accretion disk into the jet in real time. Reality TV! In this interpretation, Fig. 32 shows, at time 8.23 hr, an isolated sharp X-ray spike when accreting material is dumped into the jet. Eight minutes later, it appears as an emerging jet of particles radiating first in the infrared and then, after another seven minutes of cooling, in the radio. These data, reported by Felix Mirabel, appear to provide direct views of the disk-jet connection.

The material in the accretion disk not entering the jet would continue its descent into the black hole. The power in the jet is many million times the sun's luminosity. This is from an object no more massive than 10–15 times that of the sun! RXTE peered right down the throat of a belching black hole!!

Relativistic disk modeling and RXTE continuum spectra made possible an estimate by Jeffery McClintock and his coauthors[15] of the spin of the black hole in the GRS 1915+105 system. They found it to be at or near maximal spin, i.e., an extreme Kerr

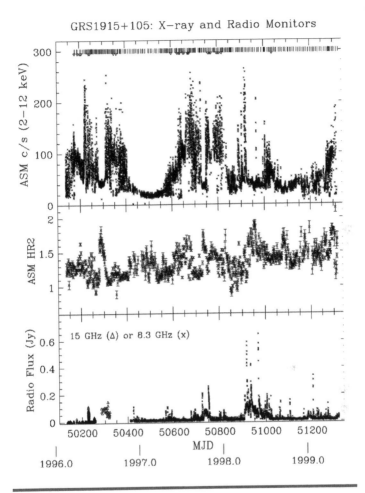

Figure 30: Microquasar four-year light curves

Top: X-ray intensity of the microquasar GRS 1915+105 as measured by the ASM over 4+ years, as a function of time (Modified Julian Days). Middle: The associated X-ray spectral hardness), the ratio of the counts in two ASM energy channels. Bottom: Concurrent radio intensity at frequency 8.3 GHz. The tic marks at the top indicate the times that RXTE was maneuvered to acquire PCA data from GRS 1915+105, one of the most wildly variable X-ray sources known. It has a number of different X-ray states, each with its type of extreme variability (Fig. 31), spectral hardness, and radio behavior. [Courtesy of Ronald Remillard; see Muno et al., ApJ, 556, 515 (2001)]

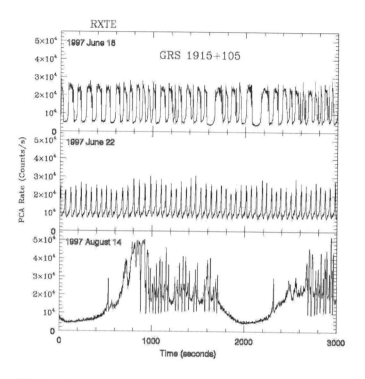

Figure 31: Microquasar rapid variability modes
Three modes of oscillation over 50 minutes of the X-ray flux from microquasar GRS 1915+105 obtained with the RXTE/PCA in 1997 on three different dates. The lower plot may represent the repetitive (~2000 s) dumping of material from the accretion disk into an ejected jet; see Fig. 32. [Courtesy of R. Remillard; see also R. Remillard and J. McClintock, ARAA 44, 49 (2006)]

black hole. In contrast, Steiner et al. found, with similar methods, the black-hole spin in LMC X-3 to be quite low.[16] The coupling between the accretion disk and the jet in GRS 1915+105 system was modeled by Rob Fender and Tomaso Belloni.[17] They made extensive use of RXTE data from this system. Their model could apply more generally to other accreting black-hole systems.

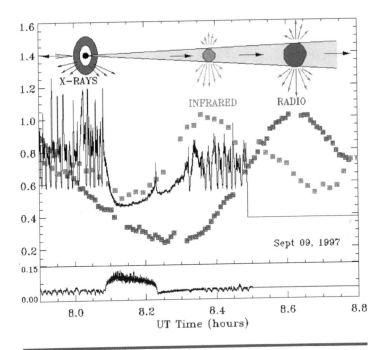

Figure 32: Disk-jet interaction in microquasar
Intensity of X-rays (RXTE PCA), infrared, and radio from GRS 1915+105 as a function of time over about one hour with the X-ray spectral hardness ratio shown below. This shows one cycle of the same dramatic repetitive pattern as seen three weeks earlier (bottom panel of Fig. 31). The rapid oscillations at left are followed by a rapid hardening of the X-ray spectrum and a subsequent softening at 8.23 h coincident with a sharp spike in X-ray intensity, which may indicate the initiating event when the disk material is dumped or sucked into the jet stream. Infrared radiation peaking eight minutes after the spike is followed by a peak in the radio flux, consistent with expanding and cooling jet material. [Mirabel et al., A&A 330, L9 (1998)]

Results such as these often require testing with further observations and modeling before one can be totally confident in the interpretations. Nevertheless, they illustrate how RXTE brought about substantial progress in the areas of black-hole spin and disk-jet coupling.

VIII
Beyond RXTE

THE CONTRIBUTIONS OF RXTE to these and other areas of high-energy astronomy are now being built upon by subsequent missions. The space agency of India (ISRO) launched a mission called *Astrosat* in 2015 with three instruments comparable to the three on RXTE but with modified technologies, except for its sky monitor, which was a near copy of the RXTE ASM. It also carried instruments for ultraviolet and soft X-ray detection. A wide range of results have been reported by Indian astronomers. US Astronomers have had little or no access to *Astrosat* data.

In 2017, the *Neutron Star Interior Composition Explorer* (NICER), a US X-ray astronomy instrument, was affixed to the *International Space Station*. It pushes beyond RXTE capability in timing and spectral resolution, while covering the soft X-ray band, where accretion disks are generally more visible. NICER's X-ray timing instrument consists of 56 small cameras, each with a concentrator and a silicon drift detector. (MIT scientists and engineers played a major role in the design, assembly, and calibration of the instrument.) As its name implies, it will use these capabilities to help deduce the equation of state of neutron stars, which determines their internal structure. Interesting results at this writing include the detailed tracking of an X-ray burst due to a thermonuclear explosion on a neutron star and the mapping of the X-ray hot spots on the surface of a neutron-star X-ray pulsar.

There has been an international effort to create a mission with the same objectives as RXTE, but with capabilities well beyond it in collecting area and spectral resolution, to be obtained by means of new technologies. An unsuccessful US proposal called AXTAR in 2010 led directly into US participation in an international proposal called LOFT, which was submitted to the European Space Agency in 2013. Sadly, it was not selected.

Currently a US-centered international collaboration continues the effort with a mission concept known as STROBE-X. Its Large Area Detector exceeds RXTE capabilities with a tenfold improvement in both collecting area and spectral resolution, using technologies developed for LOFT and NICER. STROBE-X also carries a Wide-Field Monitor that serves the sky-monitor function and a soft-X-ray Concentrator Array similar to that of NICER. The proposed mission is truly a powerhouse.

STROBE-X will address a diverse set of science goals, many of which are inspired by the great progress made by RXTE. Prime among them are the determination of black-hole spins, the equations of state of neutron stars, and the studies of phenomena in concert with other observatories, in particular gravitational-wave arrays. It has been presented as a NASA

White Paper by Paul Ray of the Naval Research Laboratory (and 132 others!) to the National Academy of Science's Decadal Survey on Astronomy and Astrophysics 2020 as a Probe class mission. It will hopefully be strongly endorsed in that survey.

It is gratifying to see the combination of RXTE features for which we fought so hard now serving as a model for observations of the highly variable X-ray sky: namely large primary detectors with a narrow field of view, a sky monitor, broad-band response, millisecond timing, rapid reorientation to any point on the sky, and a near continuous telemetry-command link.

IX
Reflections

RXTE PUT A MICROSCOPE on the regions of strong gravity and high temperatures near black holes and neutrons stars. Our knowledge of these regions and of the objects themselves, their masses, structure, angular momenta, and evolution have been greatly enhanced by RXTE observations. Other phenomena not yet understood have been revealed to humankind. All the effort that went into creating the mission proved to be well worthwhile.

Was this success the product of the skill of the RXTE engineers and scientists or simply good fortune? I think it was a lot of the former and some of the latter. Another major factor was that RXTE explored large areas of astrophysical space that had not been explored previously. In the history of science, that has often yielded major advances of understanding.

As for the MIT instruments, I have featured the dramatic near-misses we experienced and have hardly mentioned the many aspects of the ASM that proceeded with little angst that also were essential to the instrument's success: the rotation mechanism and angular readouts, the front-end electronics, the mask coding and design, the operations plan (stepping instead of scanning), the calibration of the ASM detectors (pulse height ratio to position), and finally the analysis algorithms. The EDS, as noted, was free of conceptual design difficulties in spite of the pre-launch major board failure and the DSP chip misunderstanding. The design and fabrication of the (second) flight unit led to a flawless performance for the entire 16 years.

This of course tells only the MIT part of the story. Similar teams worked hard on the Goddard and UCSD experiments, on the spacecraft, and at Goddard's RXTE Science Operations Center and HEASARC. Everyone had their technical problems but delivered successful systems on schedule and on budget also. Their stories would surely be equally exciting and revealing.

RXTE was all about people. I was impressed with how the three instrument teams worked together in resolving issues. It was crucial that scientists talked to and listened to engineers and that the engineers reciprocated. Likewise, scientists talked to and listened to NASA officials, and the officials reciprocated. Important also were the roles played by supportive letter writers, by the critics who forced us to sharpen our arguments, by the peer committees that reviewed and endorsed the mission, by theorists (especially Fred Lamb and David Pines), and by those who provided us with sound scientific arguments for the mission, even as the political and scientific landscape changed over the decades. Finally, the entire effort was made worthwhile by the army of smart observers—supported by NASA infrastructure—who proceeded to extract great science about compact objects from RXTE data during its 16 years of flight operations.

I continue to this day to be in awe of the amount of effort and talent that went into RXTE and the quality of its scientific output. I am both proud and humbled to realize I played a part in it.

ACKNOWLEDGMENTS

This book would not exist had not Dimitri Psaltis invited me to give a talk about RXTE at the Tucson workshop in 2010, and that would not have occurred had Fred Lamb not chosen to devote so much of his career to RXTE science, for which many of us are most grateful. Also necessary were the others choosing to honor Fred on his 65th birthday and also Dimitri's insistence that I write up the talk. But the story content is centered on the experiences of the RXTE team at MIT, primarily Alan Levine, Ron Remillard, Edward Morgan, William Mayer, and Robert Goeke, together with their associates. I am deeply grateful to them for their contributions to the mission and also for their memories that contributed to this work.

Drafts of this document have been read, in part or wholly, by David Helfand, Stephen Holt, Fred Lamb, Charlie Pellerin, Dimitrios Psaltis, Richard Rothschild, Dale Schulz, and Jean Swank, as well as by MIT colleagues Deepto Chakrabarty, Robert Goeke, Alan Levine, William Mayer, Edward Morgan, and Ronald Remillard. Their comments have been helpful in clarifying aspects of this story, and their tolerance of my candor in describing our interactions is much appreciated. Fred Lamb generously dug into his files numerous times to find documents and the dates of some of the critical meetings. Jean Swank and Rick Rothschild did likewise. Tod Strohmayer, Frank Marshall, and Craig Markwardt at Goddard Space Flight Center responded generously to my queries.

Alan Levine, Ron Remillard, and Wei Cui carried out the ASM science activities preparatory to launch and came up with many insights during its development that contributed to the overall success. Ed Morgan's continuous oversight of the EDS was critical in keeping it on track. The early contributions of Rodger Doxsey, John Doty and Garrett Jernigan were pivotal as recounted. Other MIT scientists, including Deepto Chakrabarty, George Clark, Walter Lewin, and Saul Rappaport proposed and carried out successful observations with RXTE. Graduate students on the program in our group at various times included Dacheng Lin, Michael Muno, Robert Shirey, Donald A. Smith, and Linqing Wen.

On the management and engineering side, a professional team led by Bill Mayer and Bob Goeke produced two fine instruments, the ASM and the EDS. Engineers Dorothy Gordon, Peter and John Tappan, and Mike Doucette were key players, as were programmers Jim Francis and Ann Davis and project technician James O'Connor.

More broadly, we, the MIT team, are grateful to the scientists and managers at NASA Headquarters, NASA's Goddard Space Flight Center, and the University of California at San Diego for their roles in the program. At MIT, the staff of the Center for Space Research, now the MIT Kavli Institute, were our bedrock of support. Finally, we all thank the taxpayers of the USA for the funds that made it all possible.

I am personally most grateful to the editor of this work, Frances King of HistoryKeep, whose contributions materially improved it, but the final edits were by me, so residual errors should be attributed to me. The designer of the print book and the covers, Lisa Carta of Lisa Carta Design, made it into a work the RXTE team can long appreciate.

.

APPENDIX I:
ACRONYMS AND ABBREVIATIONS

3A	Third *Ariel-V* Catalog
4U	Fourth *Uhuru* Catalog
A&A	*Astronomy and Astrophysics*
ACE	*Advanced Composition Explorer*
AMS	Alpha Magnetic Spectrometer
AO	Announcement of Opportunity
ApJ	*Astrophysical Journal*
ARAA	*Annual Reviews of Astronomy and Astrophysics*
AS&E	American Science & Engineering Co.
ASM	All-Sky Monitor
ATREX	Astrophysical Transient Explorer (proposed only)
AXAF	*Advanced X-ray Astrophysics Facility* (later *Chandra*)
AXTAR	Advanced X-ray Timing Array (proposed only)
BAAS	*Bulletin of the American Astronomical Society*
BATSE	Burst and Transient Source Experiment
COBE	*Cosmic Background Explorer*
COSPAR	Committee on Space Research

CSAA	Committee on Space Astronomy and Astrophysics
DSP	Digital Signal Processor
EA	Event Analyzer
EDS	Experiment Data System
EUVE	*Extreme Ultraviolet Explorer*
EXOSAT	*European X-ray Observatory Satellite*
FSS	Flight Support System
GRS	*Granat* X-ray Source
GSFC	Goddard Space Flight Center (aka Goddard)
GX	Galactic X-ray source
HEAMOWG	High Energy Astronomy Mission Operations Working Group
HEAO	*High Energy Astronomy Observatory*
HEASARC	High Energy Astrophysics Science Archive Research Center
HEXTE	High Energy X-ray Timing Experiment
IRAS	*Infrared Astronomical Satellite*
KSC	Kennedy Space Center
LAXTE	Large Area X-ray Timing Experiment
LMC	Large Magellanic Cloud
LMXB	Low-Mass X-ray Binary
LND	LND, Inc.
LSE	Laboratory for Space Experiments
MIT	Massachusetts Institute of Technology
MJD	Modified Julian Date
MMS	Multi-mission Modular Spacecraft
MRI	Magnetic Resonance Imaging
NASA	National Aeronautics and Space Administration
NATO	North Atlantic Treaty Organization
NICER	*Neutron-star Interior Composition Explorer*

NRL	Naval Research Laboratory
OSO	*Orbiting Solar Observatory*
PCA	Proportional Counter Array
PED	Payload Equipment Deck
PERT	Program Evaluation Review Technique
PI	Principal Investigator
PM	Payload Module
QPO	Quasi-Periodic Oscillation
RFP	Request for Proposal
RMS	Remote Manipulator System
RXTE	*Rossi X-ray Timing Explorer*
SAS	*Small Astronomy Satellite*
SAO	Smithsonian Astrophysical Observatory
SAX	*Satellite per Astronomia a raggi X*
SMC	Small Magellanic Cloud
SMM	*Solar Maximum Mission*
SSB	Space Sciences Board
STROBE-X	Spectroscopic Time-Resolving Observatory for Broadband Energy X-rays
SSC	Scanning Shadow Camera
TEC	Total Estimated Cost
TDRSS	Tracking Data and Relay Satellite System
TIXTE	Timing and Imaging X-ray Transient Explorer (proposed only)
UCSD	University of California San Diego
VIP	Very Important Person
XTE	X-ray Timing Explorer

APPENDIX II: LETTERS REGARDING XTE MISSION

George W. Clark (Professor of Physics, MIT) to:
  Frank Martin, Director of Astrophysics, NASA, October
  16, 1980
  Senator Ted Kennedy, October 16, 1980
  (An identical letter was sent to Representative Tip O'Neill.)
  Herbert Bridge, Director, Center for Space Research, MIT,
  October 23, 1980
  Hans Mark, Deputy Administrator, NASA, July 16, 1981

David J. Helfand, Chairman, Astronomy Dept., Columbia
  University to:
  Charles Pellerin, Director of Astrophysics, NASA, May 9,
  1988
with forwarding letter by:
  Stephen S. Holt, Chief, Lab for High Energy Astrophysics,
  GSFC/NASA), May 26, 1988
and contemporaneous penciled comments by:
  Frederick K. Lamb, Professor of Physics, U. of Illinois at
  Urbana-Champaign)

MASSACHUSETTS INSTITUTE OF TECHNOLOGY
CENTER FOR SPACE RESEARCH
CAMBRIDGE, MASSACHUSETTS 02139

October 16, 1980

Dr. Frank Martin
Director
Astrophysics Division
NASA Headquarters
400 Maryland Avenue, S.E.
Washington, D.C. 20546

Dear Frank:

Having learned in the past that the correction of the defects in NASA's proposal and selection procedures is blocked by NASA procurement and legal officials, I have written to Senator Kennedy and Representative O'Neill about my concerns. I enclose copies of my letters for your information. I hope a little pressure from the hill will make reform possible.

Sincerely yours,

George W. Clark

GWC:seb

MASSACHUSETTS INSTITUTE OF TECHNOLOGY
CENTER FOR SPACE RESEARCH
CAMBRIDGE, MASSACHUSETTS 02139
Room 37-611

October 16, 1980

The Honorable Edward M. Kennedy
The United States Senate
Washington, D.C.

Dear Senator Kennedy:

In response to a recent Announcement of Opportunity a group of
MIT scientists and engineers under the leadership of my colleague,
Professor Hale Bradt, is about to submit to NASA a 25 million dollar
proposal to develop and operate an instrument for a satellite X-ray
observatory called the X-Ray Timing Explorer (XTE). This project will
be a key element in the U.S. program of research in X-ray astronomy
in which MIT has participated with highly successful experiments on
the satellites OSO-7 (1971-72), SAS-3 (1975-79), HEAO-1 (1977-79)
and HEAO-2 (1978-). MIT has an exceptionally able group of scientists
and engineers with extensive experience in all aspects of the XTE
mission. I believe the proposal has a good chance of success if the
peer review and final selection are done competently and fairly.

Recent experiences with NASA's procedures in proposal evaluation
and selection make me doubt that they will, in fact, be competent and
fair. The present procedures have the following three major flaws:

1) Several years ago NASA changed its peer review procedures
and now no longer permits proposers to participate in the
presentation, criticism and defense of their own and
competing proposals. At best it is extremely difficult
for committee members to comprehend the technical details
of the proposals before them and to arrive at well informed
judgments of relative merits. The situation is aggravated
by the fact that most of the experts in the particular
specialty are among the proposers and are therefore dis-
qualified from membership on the peer review committee.
As a result some recent NASA peer reviews have arrived at
wrong technical evaluations that could have been prevented
by open discussion and debate in front of the committee by
the proposers. Complaints about this problem to NASA officials
have been put off with statements about what NASA lawyers
will and will not allow.

The Honorable Edward M. Kennedy
Page 2
October 16, 1980

2) NASA also changed its procedures with regard to the perogatives of peer review committees in recording their evaluations. All proposals are placed in four categories. Proposals in Category I are good in science, meet the mission objectives, and are ready to go. In Category II they are not so good in science though they meet the mission objectives and are ready to go. Previously the committees ranked Category I and II proposals as to relative merit. NASA made the final selection from Category I, or from II if there were no I's, and the selections were made in order of merit in so far as considerations of scientific balance and costs permitted. Now NASA forbids merit ranking and is therefore free to choose from Category I proposals on quite other grounds than relative scientific merit.

3) NASA laboratories which submit proposals in competition with private institutions do not include the full cost of the participation of Civil Service employees. Given several Category I proposals, which in the case of XTE will probably include one from NASA's own Goddard Space Flight Center, NASA may choose its in-house proposal on the basis of an unfair comparison of costs.

To achieve a fair evaluation and selection I believe it is essential 1) that proposers participate in the presentations, criticisms and defenses of their own and competing proposals; 2) that the peer review committee be instructed to rank proposals as to their relative merits; 3) that proposals from government laboratories be costed on precisely the same accounting basis as those from universities, specifically including all Civil Service salaries, benefits and proper overhead charges; and 4) that results of the peer review be on the public record.

I ask your help in assuring that the XTE selection will be conducted wisely and fairly, and I ask you further to consider what measures Congress might take to ameliorate the defects in NASA policy which sometimes result in the waste of resources by selection of inferior experiments.

Sincerely yours,

George W. Clark

George W. Clark
Professor of Physics

GWC:seb

MASSACHUSETTS INSTITUTE OF TECHNOLOGY
CENTER FOR SPACE RESEARCH
CAMBRIDGE, MASSACHUSETTS 02139

MEMORANDUM

TO: Herbert S. Bridge

FROM: George W. Clark *Kwc*

DATE: October 23, 1980

RE: Call from Frank Martin about my letter
             to him of October 16 with enclosed
             copy of my letters to Kennedy and O'Neill

     Frank was angry, and took me to task for not having discussed
my complaints with him before writing to the hill. He said 1) that
NASA policy is not necessarily against presentations and rebuttals;
2) in-house proposals are fairly evaluated with respect to Civil
Service costs; and 3) government procurement officials are pushing
NASA toward RFP's for defined scientific instruments instead of AO's.
He wants to discuss these matters with me in detail in the near
future, perhaps when we meet to discuss the SAS-3 data analysis.

GWC:seb

cc: H. Bradt
     C. Canizares
     W.H.G. Lewin

MASSACHUSETTS INSTITUTE OF TECHNOLOGY
CENTER FOR SPACE RESEARCH
CAMBRIDGE, MASSACHUSETTS 02139

cc: Bradt
Bridge
Canizares

July 16, 1981

Dr. Hans Mark
Deputy Administrator
Office of the Administrator
NASA Headquarters
400 Maryland Avenue, S.W.
Washington, D.C. 20546

Dear Hans:

Congratulations on your appointment. With a pioneer in X-ray astronomy at the helm NASA can't go wrong. And I think Frank Martin would be an excellent Associate Administrator for Space Science. He is a strong manager and a smart scientist with a clear sense of scientific values and political realities. I have talked with many who share this view.

Among the many urgent questions before you must be:

Whither the Explorer program? and
Which Explorers next?

The revolutionary impact that Explorer-class satellites have had on astronomy is evident throughout the literature (e.g., the Uhuru Catalog was the most frequently cited of all scientific references in 1974). But do they have a future after IRAS? The list of highly promising candidates for the new astronomy Explorers, and the enthusiastic support of the program by the community as expressed in the Field report suggest they do. Strong cases are made for Explorers in other disciplines. Their record of scientific yield per dollar is good. What they need is a shot in the arm with a funding augmentation to a level of $80M per year, and strong guidance from Headquarters that will keep their average cost below one annual program budget per mission.

As to specific Explorers, I want to argue the merits of the X-Ray Timing Explorer (XTE) and urge that it be started in time to get it into orbit during 1986.

The XTE mission, with its focus on the physical processes of compact X-ray sources, exemplifies the unity of physics and astronomy. The Pines/Lamb workshop report ("Compact Galactic X-Ray Sources") shows how XTE studies of neutron stars and white dwarfs will provide otherwise inaccessible information about matter under extreme conditions. It will lead to new understanding of superdense hadronic matter, hadron superfluidity, high temperature plasmas in superstrong magnetic fields and in gravitational fields strong enough to manifest macroscopic general relativistic effects [see Pines, Science 207, 597 (1980)].

Dr. Hans Mark
Page 2
July 16, 1981

With capabilities far greater than those of previous missions, the XTE will also address questions at the frontiers of astrophysics: the masses of neutron stars, the internal structure of normal stars (revealed through the apsidal motion of their neutron star companions), thermonuclear bursts on the surface of neutron stars, the evolution of close binaries, the sizes (from variability timescales) and mechanisms of quasars and active galactic nuclei. The XTE has a greater chance than any other conceivable mission of establishing the existence of black holes, whose clearest signatures are almost certainly to be found in the character of their millisecond variations.

The XTE will have very great scientific leverage. Its results will feed directly the mature theoretical efforts in neutron star physics, stimulate and complement investigations of optical astronomers concerned with the evolution of close binaries, and delve into new domains of short timescale phenomena where important discoveries are likely to be made. The XTE will exploit the remarkable circumstance, revealed by previous X-ray missions, that the variabilities of celestial X-ray sources are rich in physical information, a circumstance which permits a relatively simple instrument of straightforward design to collect a wealth of significant data. In the specific area of pulsating X-ray binaries, Doppler analysis of phase-coherent pulse timing measurements over periods of days and weeks yields dynamical results accurate to parts per million or better.

Since the mission will employ only proven and highly reliable technology, the chances of overruns in a strongly managed project are negligible. And the mission is well suited to serve the scientific needs of many progressive and productive astronomers whose involvement will assure a high yield of important results.

The High Energy Astrophysics Panel of the Astronomy Survey Committee, which I chaired, strongly endorsed the XTE, and the final ASC report will do likewise. I urge you to give it your careful consideration as you set the new priorities of the agency.

Sincerely yours,

George

George W. Clark

GWC:seb

196

Fred

May 26, 1988

Dear Colleague:

I enclose a copy of the letter from David Helfand to Charlie Pellerin in order that you appreciate the precise nature of the problem that we face in persuading the CSAA to reindorse XTE.

David's points are well considered; I do not think that they can be "refuted" -- rather, I think that we must present a few important scientific problems that will benefit markedly (if not uniquely or completely) from the application of XTE.

To remind you, there is a total of 1 hour devoted to our presentation at CSAA. I expect to take no more than 15 minutes on introductory information of all kinds (program organization, sensitivities, descoping, comparison with other missions, history, wish lists, cards and letters, etc -- I expect to ask HEAMOWG to help in recommending exactly what mix of material to present). Each of the guest scientists (F. Lamb, J. McClintock, and R. Mushotzky) will have about 15 minutes to present their specific important scientific utilization of XTE. If these guest scientists can send me outlines of their preesentations before the HEAMOWG meeting on June 2, some useful suggestions for CSAA might be forthcoming. My FAX number is (301)-286-3391.

Sincerely,

Stephen S. Holt, Chief
Laboratory for High Energy Astrophysics

Distribution:
  J. Swank
  H. Bradt
  R. Rothschild
  F. Lamb
  J. McClintock
  R. Mushotzky

Columbia University in the City of New York | *New York, N.Y.* *10027*

DEPARTMENT OF ASTRONOMY

9 May 1988
DJH88–57

Dr. Charles Pellerin
441 Ferry Point Road
Annapolis, MD 21403

Dear Charlie:

At your request, I record here my thoughts on the X-ray Timing Explorer mission. I have not provided copies of this letter to anyone else, but would encourage you to distribute it to whomever you believe would find its contents useful. I wish to emphasize from the start that the following represents an honest and open expression of my views as an individual scientist with a strong interest in astrophysics. I have spent many hours discussing these issues with Hale Bradt, Jean Swank, Steve Holt, members of the CSAA, and other members of the international astronomical community. I was distressed to learn from you that there exists the impression that my serious reservations about the scientific priority of XTE are interpretted in some quarters as a hostile, secretive attack covering some mysterious second agenda. I assure you this is not the case and reiterate my willingness for you to distribute this letter however you see fit.

NASA has just completed a peer review of 45 Explorer concepts and chosen four for Phase A study including FUSE/Lyman NAE and ACE; only one of these is likely to be selected for phase B study next year. It is anticipated that another call for Explorer concepts will be released two years hence. This approach to selecting and funding Explorers – fairly frequent, competitive selections and a concerted attempt not to clog the queue – have been strongly endorsed by the CSAA in its Explorer Study and by other Advisory groups. Coupled with innovative use of the Explorer budget to support highly cost-effective collaborations such as ROSAT and SXO, I believe this policy will return the Explorer program to the enormously scientifically productive vehicle for space astrophysics that it was in the 1970's, and I support it wholeheartedly.

How does XTE fit this exemplary approach to Explorers?

XTE was conceived as a mission over 15 years ago and nearly a decade has passed since it was approved without competitive review. It is, by current estimates, now six year until launch. The mission has recently been descoped. Of the three missions still in the Explorer queue, XTE is unique in that a relatively small fraction of its total cost has been expended. Given the forgoing, it seems reasonable to ask whether or not an $80M piece of the Explorer budget over the next few years is best spent by building XTE.

An independent reason for a review comes from a brief consideration of the development of international missions in X-ray astronomy over the interval since the approval of XTE. We have seen results from HEAO-1, *Tenma, Ginga,* and EXOSAT which bear on the science XTE was designed to do, Plans for GRO, SXO, GRANAT, SAX, and SPECTRA-X are well-advanced, most will certainly fly before XTE, and each has some overlap with XTE science. A serious, detailed analysis must be undertaken to determine the size of the shrinking volume of parameter space in which XTE's capabilities remain unique.

Finally, and most fundamentally, it is appropriate to examine the science case for XTE. It is not sufficient that a number of good scientists have a strong interest in studying X-ray binary stars and the brightest AGN. It is not sufficient that we be assured that 1000 X-ray sources are detectable by XTE. It is not sufficient that higher sensitivity, higher time resolution observations of quasi-periodic oscillations will become available with the PCA. It is not sufficient that the All Sky Monitor may detect another A0620 and allow us to study such a transient in great detail. It is not sufficient that a synoptic program of observations for a dozen X-ray binaries and a score of AGN will be carried out for the first time in a systematic way over a period of two years. These may well be necessary conditions for the viability of XTE, and the science team has correctly fought to maintain these capabilities in the descoped mission. But, these reasons are not sufficient. What must be demonstrated is a compelling science case which addresses in a quantitative and convincing manner the role XTE will play in answering the fundamental questions at the forefront of modern astrophysical research.

It is beyond the scope of this letter (and the capabilities of its author) to mount an exhaustive examination of the science case for XTE in order to determine whether or not it meets necessary as well as sufficient conditions to claim an $80 million portion of the highly constrained Explorer budget. As an example of the level of critical analyses which should be brought to bear, however, I will discuss briefly one of the principle objectives of XTE, its study of X-ray binary pulsars.

In Jean Swank's *"Update on XTE Objectives"* for the CSAA in January 1988 we find the following highlights of XTE pulsar science:

   I. Binary period determination

      A. for 8 of the 25 pulsar systems in which, to date the orbital period is unknown

      B. apsidal motion for known systems

   II. Identification of pulsing sources

      A. radio pulsars

      B. compact SNRs

      C. LMXBs

      D. galactic ridge transients

   III. Measurement of pulse period and profile variations

Let us examine these points in turn.

IA. How will knowing 30% more binary periods address fundamental questions concerning binary evolution and neutron star formation? Is XTE, in its low-Earth orbit best-suited for this task? Is this question not addressable by several other missions which will precede XTE?

IB. Observations of apsidal motion in these systems requires a long time base. Ninety percent of the data is "in the can" from missions over the last 20 years. Any operating satellite could do this task.

IIA. Over 100 of the most promising (nearest, fastest, highest $\dot{E}$, etc.) radio pulsars have been searched in the 1-4 keV band by *Einstein* to limits for X-ray pulses roughly two orders of magnitude more sensitive than XTE can achieve (for Crab-like spectra). For many of these, XTE luminosity limits would be above the total spin-down energy of the neutron stars. This doesn't seem like a high priority to me.

IIB. Again, *Einstein* has searched SNRs for X-ray pulsars to more sensitive levels than XTE could achieve. While it is true that higher energy observations would be useful for heavily absorbed distant sources in the galactic plane, I would guess that SXO will do a better job on this limited number of targets.

IIC. There are theoretical reasons to think that underlying millisecond periods will be hard to see in LMXBs but that is certainly not a reason to avoid looking. Some of the *Einstein* MPC and EXOSAT data can set some fairly stringent limits here; SPECTRA-X could do spectacularly well if it works as advertised. How much better will XTE actually do?

IID. *Ginga* has already found a pulsing galactic ridge transient; it and other missions could find more. How many will be needed to define the properties of this class and its place in binary evolution? Are these, as I suspect, Be star binaries? We know the population statistics of such sources reasonably well and will know it better by 1994. What will XTE add?

III. In the area of synoptic programs to study pulsar period and shape changes to learn about neutron star structure, I speak from some experience, having spent seven years as an undergraduate and graduate student carrying out such a program for radio pulsars. In those clean, isolated systems, we have learned a considerable amount (or at least have some fairly convincing theoretical models) about the interior structure of neutron stars, the interaction of crust and superfluid, etc. Glitches seem to be relatively well-understood, timing noise, less so. In this field, pulse peirods are known to 13 decimal places, period derivatives to six or seven significant figures, and time-of-arrival residuals from a best fit model are now under $1\mu s$ in the millisecond pulsars. In the X-ray binary pulsar case, these values are much more poorly determined. While I am aware of the argument that the variable torque supplied by the accreting material provides an advantage over the isolated pulsar case, offering an opportunity to record the response of the star to external forcing, I remain unconvinced that the added complications of this new *unknown* (and only indirectly measurable) factor don't outweight its advantages. I am aware of the elegant techniques developed by Fred Lamb, Paul Boynton and others to study these complex systems, but my conclusion from observing this subject over nearly twenty years is that most of what has been learned about neutron star structure has come from radio, not X-ray data.

> The study of the accretion process itself is clearly important but the aspects specific to the 25 or so X-ray binaries in our Galaxy is less so. They are not a dominant population in any energetic or evolutionary sense, and I fear, that, as the source of some of the most spectacular early results in the field of X-ray astronomy, their importance in the overall scheme of things is exaggerated by the relatively small gorup of people which has been working on them for many years. This is in no way to detract from the intrinsic interest of X-ray binaries and my view obviously includes an unavoidable element of subjective taste. However, as noted above, it is fair to ask that the few, precious missions we will see in the remainder of this century address questions of the broadest astrophysical interest. Weather in X-ray binary accretion disks and magnetospheres is, to me, not one of them.

The foregoing quick and incomplete analysis is meant not to damn the mission but to draw attention to what I percieve as a need to consider very carefully and dispassionately whether or not XTE should proceed. By raising the issue in the CSAA, I have sought to extract such a reasoned approach to this very important decision.

I have been told that I should "represent" the interests of X-ray astronomy in the CSAA and, therefore, should be supporting XTE. As this was not a condition of my appointment to the Committee, I have chosen to adopt a broader agenda, that of trying to maximize the benefit to astronomy and astrophysics of the NASA space science program. I have heard it suggested that I have some obsession with "killing XTE," born of some unsuspected pereived injustice of the past. Those on the CSAA are certainly aware, however, that this is not the only aspect of the Explorer program on which I have seen fit to comment over the past two years. It is my view, for example, that the exclusion of American participation in the analysis of EUV data from the British instrument on ROSAT, an exclusion, I am told which was designed to sustain US scientists' interest in EUVE, is an outrageous, perverse, and cynical triumph of short-term political muscle over scientific common sense. COBE has become enormously costly and represents a very serious burden on the Explorer budget; the mission may or may not yield results in proportion to its cost. These issues are, however, largely irrelevant to the case of XTE which should be decided on its merits.

It is my assumption that the XTE program will, in the end, proceed as planned. It is my hope that the mission will produce important new results on the physics of compact objects and active galactic nuclei, marking it as a worthy successor to UHURU, IUE, and IRAS. It is my considered opinon that, in the current matrix of international programs, this mission does not have an extremely high scientific priority. As a member of the scientific community, and in particular, as a member of the CSAA, I have felt it appropriate to express this opinion. I trust that my colleagues working on XTE can respect our differences of opinion, and that you will consider the forgoing as but one of many inputs into the decision-making process on the future of the Explorer program.

I would be pleased to discuss my views with you, members of your staff, members of the XTE team, or anyone else who wishes to do so. Thank you for your interest.

Sincerely,

David J. Helfand
Chair, Department of
Astronomy

Fred Lamb
May 1988 notes regarding the
Helfand Letter:

Setting up straw-men.

Representing himself, not the community.

Told what wasn't sufficient, but not is.
(No way to meet his demands.)

XTE did have backing of large fraction of
the X-ray community.

He is opinion that NS's & BH's are of little
interest is a minority opinion.

LMXBs are prototypes for AGNs, QSOs.

ENDNOTES

[1] Bradt, Swank, and Rothschild, *Astronomy and Astrophysics, Supplement Series* 97, 355 (1993)

[2] Levine et al., *Astrophysics Journal* 469, L33, (1996)

[3] The proposal was actually to search for fluorescent X-rays from the moon, which were expected from excitation due to impinging solar X-rays. Since the detectors were in a spinning rocket, a side benefit would be a search for X-rays from a large portion of the entire sky.

[4] Giacconi et al., *Physical Review Letters* 9, 439 (1962)

[5] The understanding that Sco X-1 and many other X-ray sources are *binary star systems* containing a neutron star (or sometimes a black hole) was not realized until the launch of the *Uhuru* X-ray astronomy satellite some years later in 1970; in the meantime, neutron stars were discovered as *radio pulsars* in 1967.

[6] Announcements of Opportunity apparently rested on questionable legal grounds because NASA would sole-source the delivery of the (selected) proposed instrument to the proposing university without allowing other organizations to compete for it.

7   I chuckle now reading "Nobel-winning discovery" here, which was written long before the announcement of the 2020 Nobel Prize in Physics being shared by Andrea Ghez for black-hole studies. As a member of our small group, she wrote her BS thesis on XTE EDS software in 1987. XTE may have helped spawn her interest in black holes, but it had no direct role in her great accomplishments.

8   Michiel van der Klis would surely have been included in this or another Rossi Prize for his work on kilohertz oscillations with RXTE if he had not been awarded it in 1987 for similar earlier work. In 2004, he was awarded the most prestigious Spinoza Prize in The Netherlands, based in part on his work with RXTE.

9   Van der Klis, *Annual Reviews of Astronomy and Astrophysics* 38, 717 (2000)

10   Remillard and McClintock, *Annual Reviews of Astronomy and Astrophysics* 44, 49 (2006)

11   The naming of X-ray sources is not at all systemized; there are many naming conventions. The prefixes may refer to a satellite that discovered the source, e.g., XTE J1550–564, or a catalog created by a team working with data from a particular satellite, e.g., 4U, for the fourth *Uhuru* catalog. The numbers usually refer to the celestial coordinates of the source, or less frequently the galactic coordinates. Early discoverers named X-ray sources after the constellations they were in, e.g., Sco X-1 or Cyg X-1. Sources detected by more than one satellite and listed in their catalogs could thereby gain multiple names. Old timers usually hark back to the original discovery names when referring to them. They evoke a welcome sense of the history of the field.

12   van der Klis et al., *Astrophysical Journal* 469, L1 (1996)

13   Strohmayer et al., ibid, L9

14   One might think one could resolve the 3-ms pulsing in Fig.

23 by using smaller time bins in the histogram, so that each time bin is a fraction of a millisecond. But since at most only a few tens of X-rays are detected by the PCA each millisecond, the 3–10% pulsing might not be immediately apparent. The Fourier analysis used for the contours in Fig. 23 appears to average over many pulses, perhaps about 150 of them in 1/2 second.

[15] McClintock et al., *Astrophysical Journal* 652, 518 (2006)

[16] Steiner et al., *Astrophysical Journal* 793, L29 (2014)

[17] Fender and Belloni, *Annual Reviews of Astronomy and Astrophysics* 432, 317 (2004)

INDEX

Names only.
For content, see Table of Contents and List of Figures

Index